Food, Justice, and Animals

Food, Justice, and Animals

Feeding the World Respectfully

Josh Milburn

Great Clarendon Street, Oxford, OX2 6DP,
United Kingdom

Oxford University Press is a department of the University of Oxford.
It furthers the University's objective of excellence in research, scholarship,
and education by publishing worldwide. Oxford is a registered trade mark of
Oxford University Press in the UK and in certain other countries

© Josh Milburn 2023

The moral rights of the author have been asserted

All rights reserved. No part of this publication may be reproduced, stored in
a retrieval system, or transmitted, in any form or by any means, without the
prior permission in writing of Oxford University Press, or as expressly permitted
by law, by licence or under terms agreed with the appropriate reprographics
rights organization. Enquiries concerning reproduction outside the scope of the
above should be sent to the Rights Department, Oxford University Press, at the
address above

You must not circulate this work in any other form
and you must impose this same condition on any acquirer

Published in the United States of America by Oxford University Press
198 Madison Avenue, New York, NY 10016, United States of America

British Library Cataloguing in Publication Data

Data available

Library of Congress Control Number: 2023900813

ISBN 978–0–19–286746–9

DOI: 10.1093/oso/9780192867469.001.0001

Printed and bound in the UK by
Clays Ltd, Elcograf S.p.A.

Links to third party websites are provided by Oxford in good faith and
for information only. Oxford disclaims any responsibility for the materials
contained in any third party website referenced in this work.

Acknowledgements

I wrote this book from 2019 to 2022, while I was a British Academy Postdoctoral Fellow. I thank the British Academy and my host institutions: the University of Sheffield and Loughborough University.

Particular thanks are owed to Alasdair Cochrane, whose support throughout the planning and writing of this book was invaluable. I devised and refined the project in conversation with Alasdair, and he offered comments on drafts of every chapter.

Reviewers at Oxford University Press also offered extensive, valuable comments on this project. One of these reviewers was Jeff Sebo, who kindly continued to offer comments after the completion of the review process, and helped shape the book.

I presented and discussed ideas from the book in multiple outlets, including a range of presentations at the University of Sheffield; invited lectures at Aberystwyth University and the University of Winchester; multiple presentations at the MANCEPT Workshops in Manchester; and presentations at the conferences of the Australasian Animal Studies Association, the European Association for Critical Animal Studies, and The Vegan Society. I thank audiences for—often critical!—comments and discussion on these occasions.

Ideas in the book, however, stretch back a decade to when I first seriously started thinking about food and food systems in my postgraduate studies. It is thus impossible to name everybody to whom I owe thanks in addition to those above, but among them are David Archard and Jeremy Watkins, my doctoral supervisors at Queen's University Belfast; Sue Donaldson and Will Kymlicka, with whom I worked on a previous postdoctoral project at the unrelated Queen's University in Kingston, Ontario; Bob Fischer, with whom I have co-written multiple papers on the ethics of eating animal products; and Matteo Bonotti, who first encouraged me to think about food policy and offered helpful advice on this project.

I must also acknowledge the ongoing support of my fiancée Becca, who champions my work if not always my conclusions, and of my dogs Hollie and Casper, who made the years of the COVID-19 pandemic, while I wrote this book, a bit brighter.

I dedicate this book to Siobhan O'Sullivan, who was diagnosed with ovarian cancer in 2020. Siobhan is a champion of animals, and a believer in the power of academic work to make the world better for them. Her *Animals, Equality, and Democracy* (2011) was a pathbreaking extension of liberal political thought to animal protection, and I am proud to think of *Food, Justice, and Animals* as firmly in the same tradition.

Contents

Introduction	1
1. The trouble with veganism	18
2. Bugs and bivalves	40
3. Plant-based meat	62
4. A defence of cellular agriculture	87
5. A positive case for cultivated meat	111
6. Eggs	136
7. Creating and sustaining just food systems	160
Conclusion: Having our cow, and eating her too	184
Bibliography	193
Index	204

Introduction

Imagine the burger bar of a better tomorrow—of 2030, or 2050, or 2100, or 2500. What's on the menu? Maybe we can imagine widespread changes to make a *more accessible* food system. The menu could feature the burgers, fries, and ice cream of today, but with much lower price tags. Or we might imagine widespread changes to make a *healthier* food system. Maybe the burgers are replaced with lean chicken, the fries with salad, the ice cream with fruit. Or we might imagine widespread changes to make a more *environmentally sustainable* food system. Maybe beef is off the menu, but sustainably farmed fish remains.

But we could also imagine widespread changes to make a more *animal-friendly* food system, or even one that respects animals' *rights*. What would be on the menu then?

To support our current eating habits, tens of billions of vertebrates are slaughtered annually. Many have lived lives of suffering at human hands. Trillions of fish—wild and farmed—are killed annually. And I could inflate these numbers enormously by including *other* animals[1] killed for, or in pursuit of, food—invertebrates of all kinds, male chicks killed in the egg industry, animals hunted and trapped for food, and so on. It is the food system that faces some of the largest changes if we (as a society) are going to take animals seriously.

Suppose animals have rights. This means that there are certain things that cannot be done to them without injustice. We, collectively, should protect those rights, ideally through the power of the state. This is not a book about what rights animals have, or the reasons that they have rights. There are already lots of books about those things. Instead, this book starts with animals' rights, and asks what a state that respects them looks like. It focuses on the food system of this state, as it is (arguably) the contemporary food system that is responsible for the most egregious violations of animals' rights. If the state is concerned with protecting animals' rights, that raises grave concerns about our food system. Indeed, it raises grave concerns about food systems

---

[1] In the interests of readability, I follow the conventional practice of using the word *animals* to refer to animals other than humans. This is imprecise, as humans are also animals.

altogether. Is there a viable alternative food system that respects animals' rights?

It might seem that the response is obvious. If justice entitles animals to legally protected rights, then our food systems will have to be plant-based. Animals may not be kept so that they can be killed and their bodies be turned into meat, or so that they can be kept alive (for a time) for their bodies to produce milk, eggs, honey, or other products that end up in food.

But I am not sure this answer is right. In fact, it is the claim of this book that our food systems do not need to be fully plant-based, even though animals are entitled to legally protected rights. We certainly cannot kill or torture animals for food, but they can still be involved in our food systems, and we can still eat animal products.

Consequently, I propose that the animal-rights-respecting state could contain a veritable cornucopia of animal products: the meat of all kinds of vertebrate and invertebrate animals; birds' eggs; mammals' milk; and a range of foodstuffs produced using invertebrates, such as honey, shellac, and cochineal. What is more, the rural spaces of this hypothetical state could still have fields of cows, sheep, chickens, goats, and other domesticated animals. Not only could there be animal protein for the hungry and those who value it, but there could be work with animals for those who want it. Not the brutal work of the slaughterhouse or factory farm, but *good* work—*humane* work. Perhaps in this state, we will have our cow, but, if we want to, we will eat her, too.

My vision of an animal-rights-respecting state is a vision of a state in which we do not kill animals or make them suffer for human purposes, from the trivial to the quite consequential. Borrowing an evocative term from Sue Donaldson and Will Kymlicka (2011)—without committing to the details of their position—I call this the *zoopolis*. The zoopolis is a state in which animals are respected as individuals, entitled to live as a part of our community—and, indeed, entitled to affect the direction of our community, through involvement with political processes. Nonetheless, and to repeat, I believe the zoopolis is a state in which animal products could—*should*—be available. How? Getting to the bottom of this question is the point of this book.

In the chapters that follow, I explore a range of possible routes to animal protein that are, might be, or could be consistent with full respect for animals' rights, and are thus the kinds of elements that may be part of an all-things-considered best food system for the zoopolis. In brief, I argue that some animals are non-sentient (or are almost certainly not sentient) and thus we owe them nothing directly—we may utilize them freely. Oysters might be an example. Others *probably* are not sentient, so we can utilize them with caution. Insects may belong in this category.

Other animals are sentient, or probably so, and so we owe them considerable respect—namely, rights. However, we could, consistent with this respect, welcome them into our communities and workplaces, and work *with* them to produce things of benefit to all. The confinement, objectification, mutilation, and slaughter of contemporary animal agriculture—even 'high-welfare', 'local', 'organic' animal agriculture—is inimical to respecting animals' rights. But working with animals to produce eggs, milk, wool, and more need not be. (Particularly important, for reasons I will make clear, is working with animals for *cells*.)

What about meat? 'Plant-based meat' promises burgers, nuggets, and even steaks culinarily indistinguishable from 'traditional', or slaughter-based, meat. Meanwhile, 'cellular agriculture' promises 'cultivated' meat—which is, in principle, *physically* indistinguishable from slaughter-based meat—with only the slightest animal input. Crucially, we will see, the practice of meat-eating is not *itself* rights-violating—even if (almost) all the ways that we currently acquire meat *are* rights-violating. And both plant-based 'animal' products and cellular agriculture stretch beyond meat. I contend that plant-based or cultivated milk, eggs, honey, and more may be available in the zoopolis.

I explore details of these aspects of hypothetical food systems over the remainder of the book, and I offer a more detailed chapter-by-chapter breakdown at the end of this introduction. For now, however, it is important to briefly situate the present study.

Animals, rights, and politics

We can roughly split animal rights scholarship in half. Core to the 'old' animal rights (Regan 1983; Francione 2000) is a grounding in moral philosophy and, crucially, a focus on veganism. Contrasting itself to 'welfarism'—mainstream in both animal activism (see Francione 1996) and ethical study of human–animal relationships (see Singer 1975)—old animal rights did not support simply making animal agriculture more humane. It supported *abolishing* animal agriculture, and human use of animals, all together. As Tom Regan put it,

> When it comes to how humans exploit animals, recognition of their rights requires abolition, not reform. Being kind to animals is not enough. Avoiding cruelty is not

enough. Whether we exploit animals to eat, to wear, to entertain us, or to learn, the truth of animal rights requires empty cages, not larger cages.

(Regan 2004a, 10)

In old animal rights, veganism is a 'moral baseline'—the bare minimum we owe animals (Francione 2012).

What is veganism? For the purposes of this book, I am interested in veganism as it relates to food. I treat 'vegan' and 'plant-based' as rough synonyms, adopting a practice-based definition, according to which veganism is about abstention from animal products, rather than political commitments (Dickstein and Dutkiewicz 2021). A vegan diet or food system is one without animal products. A vegan diet or food system may be consistent with animal rights. It may be the *only* food system or diet consistent with animal rights. But I contend that these are questions worth exploring; they are not true *by definition*.

Advocates of old animal rights made a case for veganism in *moral*, rather than *political*, terms. We can certainly exaggerate the extent to which the advocates of this old approach were apolitical (Cochrane, Garner, and O'Sullivan 2018), but they were generally expressing themselves as—and read by—moral philosophers. If advocates of this old approach did offer a political philosophy (see, e.g., Schmitz 2016), they would defend plant-based states, far removed from those envisioned in this book.

But there is also a 'new' animal rights, grounded in *political* philosophy, rather than moral philosophy. There has been, in the words of many commentators, a 'political turn' in animal ethics (Ahlhaus and Niesen 2015; Milligan 2015; Donaldson and Kymlicka 2017; Cochrane, Garner, and O'Sullivan 2018). This opens animal ethics—the normative study of human–animal relationships, the study of how we *ought* to interact with animals—to new vocabularies and concepts. But it also opens new questions. For example, animal ethicists have moved from asking individualistic questions about animals ('Should I be vegan?' 'How does a virtuous person treat their dog?' 'What is the value of an animal relative to a human?') to ask more societal, socio-political questions ('What laws should protect animals?' 'What forms of democracy are good for animals?' 'Can animals be our co-citizens?').

Theorists of this 'new' approach to animal rights do interrogate food (e.g. Donaldson and Kymlicka 2011, 134–139; Cochrane 2012, chapter 4; Wayne 2013), setting up in opposition to the 'abolitionism' of old animal rights. Veganism ceases to be a moral baseline, and respect for animal rights ceases to mean the end of human–animal interactions. Consequently, theorists of new animal rights are explicitly open to non-vegan diets and states, but their comments stop short of worked-out food systems. Perhaps the envisioned production only works on a small scale—individuals or small communities—and thus fall short of being a contender for a food *system*. Alternatively, or additionally, comments on food production may be brief, inconclusive, or scattered.

Take, for instance, Donaldson and Kymlicka. We can put aside the details of their framework for now—let us simply ask what they say about the food system of the zoopolis. The answer is: surprisingly little. Like all advocates of animal rights, old or new, they see contemporary food systems as beyond the pale. Recognizing animals' rights, they say, means recognizing that animals were not put on the earth to feed us, while the rights themselves rule 'out virtually all existing practices of the animal-use industries' (Donaldson and Kymlicka 2011, 40). They consequently speak in glowing terms of a 'switch to veganism ... on a societal scale' (2011, 202).

This rejection of animal foods leads Robert Garner, another advocate of new animal rights, to lament a missed opportunity for novelty in Donaldson and Kymlicka's work. For him, their 'starting point is the acceptance, as a baseline, of a traditional species-egalitarian abolitionist animal rights agenda' (Garner 2013, 103)—the old approach to animal rights against which Donaldson and Kymlicka position themselves. Garner sees Donaldson and Kymlicka as offering a politicized vision of the old animal rights. Their framework, he says,

> has the effect of ruling out of account the domestication of animals for exploitative human purposes. Animal agriculture [is] therefore morally illegitimate on the grounds that to use animals in such a way is to infringe their rights. ... [Donaldson and Kymlicka] adopt an animal rights ethic that rules out the use of animals for food[.]
>
> (Garner 2013, 104)

But Garner is only half right. For Donaldson and Kymlicka, co-living with animals *can* be consistent with sharing milk with cows, or eating eggs laid by chickens, or gathering wool (for food additives) from sheep (2011, 136–139). And they ask questions about both the corpses of animals who have died naturally or accidentally, and about cellular agriculture—two potentially rights-respecting sources of meat (2011, 151–152). They do not raise some other questions. For example, they do not address (so do not rule out) humans eating non-sentient animals,[2] and they do not address cellular agriculture beyond cultivated meat.

There are, however, at least three reasons to question Donaldson and Kymlicka's scattered comments on using animals for food. And, to repeat, my exploration of Donaldson and Kymlicka's comments is indicative. I could level similar criticism at other scholars of new animal rights.

First, Donaldson and Kymlicka hedge their arguments, reaching no firm conclusions about animal products. While they do not rule animal products out, neither do they really endorse them.

Second, Donaldson and Kymlicka do not explore the reasons that we might have for endorsing non-vegan food systems, even as advocates of animal rights. To be clear, there are such reasons. Garner (2013) offers arguments about the *practical* benefits for animals and the animal-rights movement; his reasons are pragmatic. My own reasons, explored in Chapter 1, are principled.

Third, and perhaps most surprisingly, Donaldson and Kymlicka's discussions of animal products are noticeably small-scale. Their discussions evoke images of, for example, families keeping chickens in backyards and collecting (though not selling) eggs, or individuals collecting the corpses of animals killed on the road to feed companion cats. There seems to be little by way of a food *system* emerging from these ideas.

Thus, on how citizens of the zoopolis eat, Donaldson and Kymlicka are caught between the straightforward—abolitionist—vision of the vegan state, and *something else*. This something else is a food system far from those in twenty-first century states (or any historical state), but it is unspecified, or, at least, *under*specified. It is the goal of this book to specify that 'something else', and offer a vision of a non-vegan food system that might sustain the zoopolis. Perhaps that zoopolis resembles Donaldson and Kymlicka's vision, but perhaps it resembles one of the (more or less) competing or complementary visions offered by other (more or less) liberal thinkers working in the 'political turn'. I refer to many of these thinkers' ideas over the coming chapters.

This book offers clear steps towards a food system for animal rights theory—but a food system for *new* animal rights, and a long-needed alternative to the food system of the abolitionists. But, more than this, it offers a food system that people committed to a wide range of competing values—animal advocates and animal agriculturalists; vegans and foodies; environmentalists and food-justice advocates—could champion.

---

[2] Donaldson and Kymlicka are sentientists (see Chapter 2), so they will not conceive non-sentient animals as rights bearers. To my knowledge, however, they never address this issue explicitly.

Animals and justice

For theorists of animal rights, old or new, the standard practices of western animal agriculture are unjust. With few exceptions, it is unjust to slaughter animals; confine them to cramped sheds; mutilate them. Thus, for animal-rights theorists (in contrast to some other animal ethicists) justice has always been important. However, for the theorists of new animal rights, it is central—a defining concern. As three scholars central to the political turn put it,

> the crucial unifying and distinctive feature of these contributions [to animal ethics in political philosophy]—and what can properly be said to mark them out as a 'political turn'—is the way in which they imagine how political institutions, structures and processes might be *transformed* so as to secure justice for both human and non-human animals. Put simply, the essential feature of the political turn is this *constructive* focus on justice.
> (Cochrane, Garner, and O'Sullivan 2018, 263–264)

It should be clear that the current project aims to imagine how political institutions, structures, and processes—specifically, food systems—might be transformed. But what, precisely, is *justice*?

Justice has been central to political philosophy since at least Plato's *Republic*, which is essentially an enquiry into the meaning of justice. For Plato, justice is ultimately about everything (including people) being in its (their) proper place, and thus about everything (including society as a whole) being balanced. In contemporary political philosophy, a focus on justice is owed in part to John Rawls. He distinguishes between the *concept* of justice and varied *conceptions* of justice. I will return to specific conceptions shortly, but what they share—and thus what gets to the bottom of the *concept* of justice—is that they offer:

> principles for assigning basic rights and duties and for determining what they take to be the proper distribution of the benefits and burdens of social cooperation. … Those who hold different conceptions of justice can, then, still agree that institutions are just when no arbitrary distinctions are made between persons in the assigning of basic rights and duties and when the rules determine a proper balance between competing claims to the advantages of social life. Men can agree to this description of just institutions since the notions of an arbitrary distinction and of a proper balance, which are included in the concept of justice, are left open for each to interpret according to the principles of justice that he accepts. These

principles single out which similarities and differences among persons are relevant in determining rights and duties and they specify which division of advantages is appropriate.

(Rawls 1999, 5)

When Rawls says *persons*, he is imagining humans—or perhaps *some* humans. But (because animals can be persons, because justice can stretch beyond persons, or both) we need not limit justice to humans. The point of the political turn in animal ethics is that discussions of justice in the twentieth century ignored animals—theorists relegated questions about animals to 'mere' morality. At best, questions about animals were a *later* or *derivative* question for justice, rather than something theorists must principally address (see Nussbaum 2007, chapter 1).

Justice for animals should not arise only 'after' we have resolved questions about justice for humans—in theory *or* practice. Justice is important for animals because the demands of justice are particularly stringent. They are not merely things it would be *nice* or *good* or *charitable* or *virtuous* to do. They are not even things that we should insist upon—at least, we should not *stop* at insistence. Instead, the demands of justice are so stringent that people might legitimately *coerce* others into meeting them.

For example, it might be nice, good, charitable, or virtuous of me to donate much of my income to a worthy charity. If my moral convictions lead me to endorse the effective altruism movement, or Islam's *zakah* principle,[3] I might even think I am *morally required* to donate. But it is unlikely that I think that I am *required by justice* to donate large amounts. Indeed, this sounds paradoxical; if I am required by justice to donate, it does not sound like *donation* at all. Donations are, by definition, supererogatory (that is, they go beyond my duty). If I fail to 'donate' when required by justice to do so, then others (likely acting on behalf of the state) may (or should) *force* me to donate, and perhaps punish me.

In focusing on justice, political philosophy is a high-stakes branch of ethics: 'It's not just about what people ought to do, it's about what people are morally permitted, and sometimes morally required, *to make each other do*' (Swift 2013, 24, emphasis added).

If our duties towards animals are matters of mere morality, then, no matter how important they are within our moral theory, we have no business forcing each other to obey them. In a liberal society, disagreements about

---

[3] I thank an anonymous University of Sheffield student for this example.

morality are to be tolerated. If I think alcohol is immoral, but you think teetotalism is prudish self-abnegation, then so be it. We must accept that the other is (perceived to be) wrong. You have no business forcing me to try fine wines, and I have no business stopping you from enjoying them. To borrow the words of John Stuart Mill, although noting they were originally presented in a slightly different context, my feeling that someone is not behaving in accordance with the (from my point of view, and for all anyone else knows, *true*) norms of virtuous consumption might present 'good reasons for remonstrating with him, or reasoning with him, or persuading him, or entreating him, but not for compelling him, or visiting him with any evil in case he do otherwise' (Mill 2008, 14).

It should thus be clear why relegating harms to animals to the domain of the merely moral will be very bad for animals in a liberal state (Garner 2013, chapter 3). Imagine you think it morally acceptable—even morally good—to do awful things to animals. Imagine, for example, that you are a gourmet of the most morally dubious kind, and think that it is *good* that people catch ortolan buntings, stab out their eyes, force feed them, then drown them in brandy so that they can be roasted and eaten. Or consider other ostensibly cruel foodstuffs, which you might think it *good* to experience: the foie gras of France, which involves farmers force-feeding birds to engorge their livers; the *ikizukuri* of Japan, which involves chefs cutting and serving live fish, octopuses, and others; lobsters, who cooks drop live into boiling water; or the Yin Yang fish ('dead-and-alive fish') of Taiwan, which has a deep-fried body, attached to a still-living head.

Or maybe you think it is *good* that chefs support killing animals particularly rare, or particularly intelligent, or particularly human-like, to offer unrivalled dining experiences: Central African bushmeat made from gorillas and chimpanzees; Japanese or Faroese cuisine containing dolphin or whale meat; dogs bred for the Korean dog-meat trade.

Imagine, too, that you are unmoved—to evoke Mill—by my remonstrating, reasoning, persuading, or entreating. What then? In the liberal state, if our duties to animals are not matters of justice, I must simply tolerate your egregiously harmful behaviour, just as the teetotaller and sommelier must tolerate each other.

But someone seems to have been forgotten, here. People with different visions of the good life tolerating each other's pursuit of happiness is one thing. Indeed, it is central to the current project. But when that pursuit of happiness involves someone else—the French goose, the Korean dog, the Faroese whale—questions of justice enter.

This is where it is worth reintroducing *conceptions* of justice. This book begins with the contention that animals have rights. As a conception of justice (or, more precisely, set of conceptions, or part of a conception), animal rights takes as arbitrary the idea that humans are some*ones*, about whom justice claims can be made, while animals are some*things*, the use of which must be tolerated. (A story about what constitutes an arbitrary distinction in the attribution of rights and duties, recall, is key to Rawls's explanation of conceptions of justice.) And if animals are someones, entitled to protection, it is only natural that this is put into the language of rights. Like justice, rights can be understood as those moral entitlements for which one can 'demand or enforce compliance' (Nozick 1981, 499). For political philosophy, then, rights are central. This is because political philosophy, in its focus on justice, is the 'theory of what behavior legitimately may be enforced, and of the nature of the institutional structure that stays within and supports these enforceable rights' (Nozick 1981, 503).

In this Nozickian language, my question in this book is which ways of acquiring food the state may prohibit, and, more importantly, which they may *not*, to ensure that the state does not overstep its mark. For liberals, any restriction of freedom is regrettable. We must ensure that animal rights do not become, in the words of one critic, tools that 'only confer additional power on governments and bureaucrats to run our lives for us', rather than genuine protections for animals (Machan 2004, 23). State interference with liberty must go far enough to realize justice—for humans *and* animals—but no further. If it goes further, the state will, itself, be guilty of unjust impositions. Achieving justice is a delicate business. Too little interference, and one set of individuals (here, animals) lose protections to which they are entitled. Too much, and another set of individuals (here, humans) lose liberties to which they are entitled.

But there is a puzzle here, which I must acknowledge. It is fine to say we must tolerate disagreements about morality, while we can enforce the demands of justice. But there is disagreement about what constitutes justice. For example, many liberals will disagree that animals have rights. Well, so be it—again, this is not a book arguing that animals have rights, it is a book exploring the consequences of animal rights. If my critics prefer: it is a book exploring what it would mean *if animals had rights*.

But this does not solve the problem, as there is still room for disagreement between advocates of animal rights about *which* rights animals have—or which of our duties to animals cross the line from *mere* morality to issues of justice. I hope to sidestep many of these disagreements, but I will say a little more about the rights animals have later in this introduction.

And yet there is more disagreement I must acknowledge. There is disagreement—even between liberals—about the areas over which the state has obligations (or permissions) to intervene. Put another way, there is disagreement over *how much* interference and control over our lives liberal states permissibly exercise. Again, I hope to be able to sidestep many of these concerns by not taking too strong a stance. I am liberal, but I am not concerned here with specifying exactly what 'flavour' of liberal. Nonetheless, I will return to this issue in Chapter 7.

Before moving on, let me briefly return to ortolans, dolphins, *ikizukuri*, and the rest. Strikingly, while no state recognizes animal rights as endorsed in this project, liberal states typically *do* ban many of (or all) the practices I listed earlier. If I lived in Europe or North America and I knew that my neighbour was slaughtering her dog for some Korean cooking, or chatting with an ortolan catcher about sourcing buntings, or planning to open a Japanese restaurant specializing in dolphins, whales, and *ikizukuri*, I could contact the authorities. And, depending on the laws of my state (and the priorities of the authorities …), I might be able to use the coercive power of the state to stop (or punish) my neighbour. Maybe, then, animal rights as a matter of justice are not *too* alien to contemporary liberal societies.

Food justice

It is worth mentioning another side to discussions of justice that, regrettably, has an ambivalent relationship with animal rights: *food justice*.

Like the liberal ideal of mutual toleration of pursuit of the good life, the traditional concerns of food justice are central to the motivations of this project. These include access to good, plentiful food for all; access to culturally appropriate and important foods for those of minority communities; availability of good work in the production of food and the abolition of exploitative labour practices; the production of food in ways respectful to the local and global environment (and the people who live in these places); security, resilience, and independence in food production; and popular and local control of food production and distribution. These are real issues, and reference to them will appear throughout the book. But my focus in this section is not on the issues themselves, but on the relationship between food justice and animal rights.

Food justice emerged as an overlapping concern of several activist, scholarly, and scholar-activist communities. Given this organic birth, the concept remains relatively loosely defined—or perhaps a 'work-in-progress', and one that academics can mould and influence (Gottlieb and Joshi 2010, 5–6).

Nonetheless, the food justice literature, like the literature on justice in political philosophy, is anthropocentric, overlooking the possibility that animals, too, have entitlements of justice. At best, there is the occasional more-or-less perfunctory nod towards 'animal welfare'.

In an overview, Alison Hope Alkon says that 'Food justice research explores how racial and economic inequalities manifest in the production, distribution, and consumption of food, and the ways that communities and social movements shape and are shaped by these inequalities' (2012, 295). It is undeniable that these racial and economic inequalities are matters of justice. But it is far from clear that they should be the *only* inequalities of concern in food-justice scholarship. Admittedly, Alkon's account does not preclude the possibility that concern for animals can be (or is) a focus of food-justice research—and she approvingly quotes a definition of food justice mentioning animal welfare. But, despite exploring avenues for further research in food justice, she does not suggest that scholars further incorporate animals.

Indeed, Alkon—as a co-editor of the influential *Cultivating Food Justice* (Alkon and Agyeman 2011)—shares responsibility for the anthropocentric nature of food justice research. *Cultivating Food Justice* focuses on race and class, and while several chapters discuss animal advocates, this is generally in the context of criticizing them for their inattention to racial inequalities and injustices.

In his exploration of food justice, Kyle Powys Whyte does not emphasize race and class as much as Alkon, but his approach remains human-focused. He writes that '*Food injustice* occurs when at least one human group systematically dominates one or more other human groups through their connections to and interactions with one another in local and global food systems' (Whyte 2018, 345). Again, this does not strictly imply that animals cannot be victims of food injustice. But his repetition of the word *human* perhaps indicates where his sympathies lie.

Ronald L. Sandler offers a more open definition of food justice. Defining *distributive justice*, he writes that 'A system, policy or practice is thought to be unjust if those who benefit from it do not also shoulder the associated burdens, or if the benefits of the system are unequally distributed without good reason'. Meanwhile, food justice 'concerns the allocation of benefits and burdens within the food system' (Sandler 2015, 26). His examples allude to animals, but it is (for instance) fishers who are the victims of injustice—not fish, who are mentioned only indirectly. Nonetheless, his definition *could* capture animals: Humans might force animals to shoulder the burdens of the food system for human benefit, or fail to share the benefits of the food system with animals without good reason. (Assuming that animals' interests matter, there is no *might* about it.)

If animals are rights-bearing beings who impose demands of justice on us, it is reasonable to conclude that ensuring that the food system treats them as they deserve is a matter of food justice. It is not my aim here to provide a concrete and all-encompassing definition of food justice. However, I will be using the term. Although this is not a book offering sustained analysis of race or class, it is a book that takes seriously concerns at the centre of food justice activism and scholarship, such as access to food, good work, sustainable and environmentally friendly food systems, respect for food customs and practices, and so on. It is also a book about justice, and about food.

I hope that readers perceive this book as a contribution to the food justice literature, although fear that its focus on animals will be off-putting to many advocates of food justice, who might see animal rights as a distraction from real matters of justice at best, or a colonial project at worst. These are contentions that I reject, but I reserve a full treatment of this question for another time.

Animals and rights

I must address a final matter before this enquiry can begin in earnest. I have said that this project takes for granted that animals have rights. But I have said little about them: why animals have rights, and what rights they have. It is worth saying something about both topics as a preliminary clarification—although analysing precisely which rights animals have, and what these rights mean, will be a topic that I revisit throughout. For example, when addressing plant-based meat, we must ask whether Susan Turner (2005) is right to say that animals' rights prevent us from representing them as a mere resource. If so, that could rule plant-based meat out of our rights-respecting food system—assuming, of course, plant-based meat *does* represent animals as a mere resource. But this is a conversation for later.

Following a theme in the political turn in animal ethics (Milligan 2015), I take it that animals have rights because they are the kinds of beings who have interests. A focus on interest-based rights is explicit in many works in the political turn (e.g. Cochrane 2012; Garner 2013) and implicit in much of the rest (e.g. Donaldson and Kymlicka 2011; Meijer 2019). Interest-based rights approaches hold that (some) animals have a welfare—things go better or worse for them. This is not the same way that things go better or worse for a table or tree. Instead, it is about the quality of the animal's life from the point of view of the animal—although with the caveat that this need not reduce solely to what the animal herself wants, what the animal believes is in her interest, or what makes the animal happy. Indeed, there is room for disagreement about what is in an animal's interests. I will return to this issue repeatedly.

A basic list of rights possessed by animals will include presumptive rights against being killed or made to suffer (Cochrane 2012)—and, we may add, a right to shape the direction of the community of which one is a member (Cochrane 2018). A more extensive set of rights would entail that we recognize animals' rights to freedom, and to all kinds of positive entitlements—perhaps determined by the relationships we have with them (Donaldson and Kymlicka 2011). Indeed, a focus on the different groups that animals are members of has been a key contribution of the political turn. (Claims about relationship-based rights may sound suspect, but they are perfectly familiar. I have more duties, including duties of justice, concerning *my* dogs than I have concerning yours.)

There may be circumstances in which rights are permissibly infringed—tragic cases of genuine conflict, such as the much-touted sinking lifeboat or barren desert island, may justify killing animals or humans (Francione 2000, chapter 7). But, generally, rights act as 'trumps' (Dworkin 1984) or 'side-constraints' (Nozick 1974, 29). *Trumps*, because rights trump other kinds of reasons that might speak against their respect, even when these reasons might otherwise be very good, and *side-constraints* as our actions, even when pursuing the best of outcomes, are constrained by rights' existence. They are lines that, in normal circumstances, we may not cross.

The law does not recognize these rights. It should. As established, this is a book about what a state complying with its duty to protect animal rights would look like. It is, then, a work of *ideal theory*, rather than *non-ideal theory*. Like discussions of justice in political philosophy, this distinction goes back at least to Plato (see *Republic* 473), but owes its current popularity to Rawls (1999). A large and complex literature on the distinction between ideal and non-ideal theory has emerged in the twenty-first century. It would be unhelpful to review this here. Instead, we can summarize the distinction as simply this: Ideal theory is about what a just society would look like—it is limited only by the constraints of possibility. Non-ideal theory is about making unjust societies more just, and, ultimately, transitioning unjust societies to fully just societies. It is limited by all the practical constraints impacting our political action in the given unjust societies.

Even though non-ideal theory is the important aspect of political philosophy, as it is non-ideal theory that allows us to remedy injustice in the real world, ideal theory has a certain priority over non-ideal theory—at least, that is what Rawls believes (1999, 8). This is because our ability to identify injustice and transition towards a more just society is comparatively minimal until we have an idea of what a just society looks like. Consider the present case. Whether or not the zoopolis will be a vegan state will make a difference to how animal activists can most effectively aid animals today—at least if we understand activists *effectively aiding animals* as activists bringing about a world in which we realize justice for animals.

Now, I am not claiming that if the zoopolis is not vegan, we should not engage in pro-vegan activism, while if the zoopolis is vegan, all our activism must be pro-veganism. Philosophers do not present ideal theory as practical guidance for the here and now, hence the need for non-ideal theory. But, as I explore in Chapter 1, this does not mean that ideal theory is not useful for building bridges in the real world. If animal activists and their most vehement opponents can agree on a vision of an ideal world, the battle is half over. The non-ideal consequences of the ideal theory offered here, naturally, warrant exploration. I offer the beginnings of such exploration in Chapter 7—although such comments must necessarily be preliminary.

A final note on ideal theory. Ideal theory is not utopianism—understood here in the pejorative sense of unachievable, head-in-the-cloud theorizing. Ideal theory is about coming up with principles and institutions that we *could* realize, if only we had the collective will to do so. In the gendered—and anthropocentric?—language of Jean-Jacques Rousseau, it is about 'taking men as they are and laws as they might be' (1968, 49). Even if the vision offered in this book is far from current practice, it is one that we could collectively institute—if that is what we chose.

Outline of the book

In Chapter 1, I lay out the motivation for exploring a non-vegan but animal-rights-respecting state. I show that there are questions we should raise about vegan food systems from both an animal-rights and a liberal perspective. These are *principled* problems—while there may be good pragmatic grounds for exploring non-veganism, these are not my focus.

These ideas in hand, I turn to explore industries that could be consistent with animal rights. In Chapter 2, I address the prospect of farming invertebrates, whether for their bodies or their products. At the core is the question of sentience. If these animals are sentient, they are rights-bearers, and entitled to protection. If they are not sentient, we can farm them freely.

Difficult questions emerge in the case of animals about whose sentience we are uncertain. And, of course, if we are unsure about whether animals are even sentient, getting to the bottom of what is and is not in their interests is difficult. We can not, I suggest, reach simple conclusions: While we can farm some invertebrates freely, some we must farm with caution, while others we may not farm at all.

Chapter 3 turns to the prospect of making 'animal' products out of plants, focusing on plant-based meats. While these might sound innocuous, they face criticism from advocates of veganism for supposedly reinforcing, rather than challenging, slaughter-based animal products. And they face criticism from both advocates of plant-based whole foods *and* advocates of slaughter-based meat for, in a variety of ways, being 'bad' food. I argue that we can overcome these challenges.

Many of the same challenges arise when it comes to the budding industry of 'cellular agriculture'. This cluster of technologies can produce meat, milk, and other animal products with minimal animal input. Chapter 4 explores challenges specific to the cellular agricultural technologies of cultivated meat and precision fermentation—the latter of which produces, among other things, cultivated milk. This chapter thus offers a negative case for cellular agriculture as part of the food system of the zoopolis.

Chapter 5 stays with cultivated meat, offering a *positive* vision of how cellular agriculturalists could produce meat in an animal-rights-respecting state. A central puzzle involves how cellular agriculturalists source cells. I argue that they could source cells in several permissible ways. In particular, if we can reconceive animals as *workers*, then animals *protected by workers' rights* could provide the needed cells. This possibility allows us to envision a truly ideal cellular agriculture—good for people *and* animals.

If animals can work on farms to produce cells, could they work on farms to produce other products? In principle, yes—although slaughterhouses are out of the question. As an example, Chapter 6 looks to eggs. I argue that as well as living with chickens and eating their eggs, people could source eggs from farms in which farmers treat chickens with the respect they warrant. Of course, these farms face ethical challenges—and they could not produce *cheap* eggs. But I argue that there could be a place for such farms, and that it is likely that there would be a market for their eggs.

Chapter 7 steps back from questions about particular foods in order to explore the question of bringing about and maintaining the food system that the book has pointed towards. It explores what it means for a state to *permit* and *support* particular (elements of) food systems. But it also asks what the arguments of this book mean for actors today: what they mean for individuals' dietary choices; for the campaigning of activists and activist organizations; for states.

A short concluding chapter summarizes the arguments of the book, reflecting on what the book has, and has not, done. It asks about ways in which the argument could be extended, and asks what it means to get these questions right—and, perhaps more importantly, wrong.

1
The trouble with veganism

Animal advocates should explore the possibility of non-vegan food systems in the zoopolis. Indeed, the zoopolis should (probably) permit non-vegan food production. And, further, the zoopolis should (probably) actively *endorse and support* a non-vegan food system. At least, these are the claims I motivate in this chapter.

These claims are controversial. Even the first will raise eyebrows—so let us begin there. Why would advocates of animal rights explore the possibility of non-vegan food systems when it is animal agriculture that has (arguably) been responsible for the most egregious violations of animals' rights?

Animal-based food systems are far from necessary. After all, we do not need to eat animal products to be healthy. Indeed, there is empirical evidence suggesting that plant-based food systems would be healthier. I do not mean this simply in the sense that a vegan food system could lead to lower levels of certain health problems—obesity, diabetes, certain cancers (see, e.g., Melina, Craig, and Levin 2016)—but because of the health benefits associated with vegan food systems' environmental credentials (see, e.g., Scarborough et al. 2014); with lower risk of zoonoses (Bernstein and Dutkiewicz 2021; Knight et al. 2021);[1] and with lower deployment of antibiotics, the use of which are prevalent in (intensive) animal agriculture, risking antibiotic resistance (OECD 2016).

These gains could lead to substantial economic benefits (Springmann et al. 2016; Tschofen, Azevedo, and Muller 2019). There is even evidence that a plant-based society would be safer—areas around slaughterhouses have increased violence, including sexual violence (Fitzgerald, Kalof, and Dietz 2009).

Thus, not only is a plant-based food system possible, but perhaps we should desire it for reasons beyond animal rights. If animals have rights and a plant-based system is possible, why explore a non-vegan food system at all?

---

[1] When I first wrote these words, the world had not felt the effects of COVID-19, a zoonotic disease. I believe they take on new resonance now. No longer can people plead ignorance or downplay the threat of zoonoses.

Food, Justice, and Animals. Josh Milburn, Oxford University Press. © Josh Milburn (2023).
DOI: 10.1093/oso/9780192867469.003.0002

I will note some pragmatic reasons to explore a non-vegan food system—although these are not my present focus. The bulk of the chapter will search for more principled reasons for activists and theorists—and the policymakers in the zoopolis—to explore, permit, or support non-vegan food systems.

The conception of justice I have endorsed is a liberal approach to animal rights. Separating the conception into its constituent parts, this chapter will explore both *liberal* and *animal-rights* reasons for exploring non-vegan food systems. I split liberal reasons into two groups: Reasons grounded in respecting diverse conceptions of the good, and reasons grounded in the demands of food justice (understood in anthropocentric terms). I also group concerns related to animal rights into two categories. One concerns harms to animals in arable agriculture. The other concerns the paradoxical 'extinctionist' conclusions of old (and new) animal rights.

The chapter closes by bringing together the liberal and animal-rights critiques of veganism, resulting in the three claims with which I began. Each claim builds upon the last, and each is more controversial than the last. First, I conclude that liberal theorists of animal rights should *explore* and *be open to* the possibility of non-vegan foods in the zoopolis. Second, I conclude that the zoopolis should, in principle, *permit* the production of non-vegan foods. And third, most controversially, I conclude that zoopolis should (in principle, depending upon a complex range of factors) *support* a non-vegan food system if—and only if—such a system could be consistent with respect for animal rights.

Pragmatic problems with veganism

Garner suggests that the most important thing political philosophy can bring to animal ethics is a dose of pragmatism—a move away from the putatively unachievable goals of old animal rights, and towards something realizable (Garner 2012; see also Milligan 2015; Hadley 2019a). What might this mean? A plant-based zoopolis would ban slaughtering animals for food. But advocates of animal rights could suggest that, whatever the merits of this ban *in principle*, it should not be endorsed *in practice*—at least for now (Garner 2015, 216).

In the introduction, I mentioned Garner's *pragmatic* arguments for activism for a non-vegan food system. He endorses a non-ideal legal system permitting killing animals, but not making them suffer (Garner 2013, chapter 8). Thus, farmers could raise pigs and kill them for bacon; raise chickens for eggs, killing unneeded male chicks and less productive hens; raise cattle for milk and kill bullocks and older cows. At least, they could *in theory*. When Garner expands on the practical consequences of his approach, he talks of cultivated meat and animals genetically engineered not to feel pain, worrying that even the most animal-friendly extant forms of pastoral agriculture involve suffering (2013, 137). Perhaps, then, the position he endorses is not so pragmatic—or, jumping ahead, so far from my own.

Endorsing suffering-free or suffering-*light* animal agriculture as a pragmatic strategy is compatible with many visions of the ideal, from the conservative to the radical. Animal activists supporting (say) welfare reform and anti-cruelty legislation in practice might have principled disagreements with each other. Meanwhile, activists who share a vision of the ideal might disagree about activist methods. There is thus room for discussing whether endorsing non-vegan foods could be part of a pragmatic approach to animal rights.[2] I return to these questions in Chapter 7.

Nonetheless, these pragmatic reasons are not what interest me here. Instead, I hold that we have good *principled* reasons to explore non-vegan food systems as an ideal to aim for. That said, principled cases for non-vegan food systems could themselves have pragmatic value. In short, this is a book about having our cake and eating it too—or having our *cow* and eating *her* too. It is a book about how we can have the best of both worlds; how we can have respect for animals, *and* access to the positive things that animal agriculture gives us: Good food, good jobs, and more.

Most of veganism's critics drawing upon the principled reasons for questioning veganism explored in this chapter—including Jocelyne Porcher, Jean Kazez, and Matthew Evans, who will be introduced shortly—accept that harm to animals is a bad thing. Indeed, it is difficult to find someone denying it.[3] Although their actions may lead us to question their sincerity, it is hard to find someone who does not express regret at the suffering caused by industrial animal agriculture. Even the most ardent defenders of meat express regret at what happens in slaughterhouses, and are often keen to stress this—indeed, we might wonder about anyone happy with how slaughterhouses operate.

All of this is true even if there are reasons to be worried about veganism. But what if we could have a solution that respects animals, but overcomes the worries of veganism's critics? Even if this is far from the current situation,

---

[2] I am sceptical of the claim that supporting 'high-welfare' farming is good for animals, but I have offered pragmatic arguments for permitting/supporting non-vegan foods (e.g., Milburn 2016, 2018; cf. Fischer 2020).

[3] There are contrarians, who either bend over backwards to deny that animals are *really* harmed, or else twist moral arguments until harming animals is unproblematic.

the prospect of agreement between animal advocates and voracious (but reasonable) critics of veganism is something that could have immense pragmatic value. It offers hope of a shared vision of the future, and an antidote to hostility around veganism in the public eye.

To take a few headline-grabbing examples, vitriol targeting vegans from the likes of Piers Morgan (the provocateur who called a bakery chain 'PC-ravaged clowns' for offering a vegan sausage roll) and William Sitwell (the magazine editor who joked about killing vegans in response to a proposal for articles about veganism) fan the flames of so-called 'vegaphobic' tendencies. Meanwhile, a Leeds butcher flailed at protesters with raw meat in 2018, and 'anti-vegan' protestors faced fines for eating raw squirrels at a London market in 2019. Hostility goes both ways, with animal activists sometimes opting to use decidedly uncivil approaches. French activists have been handed jail time for protests targeting farmers, butchers, and restaurateurs, with those involved in the meat industry calling for greater police protection. The image of butchers living in fear from vegan attack evokes the personal (sometimes disturbing) attacks associated with Oxford anti-vivisectionist protests over several decades.

Now, I do not deduce anything about the wider culture from these stark examples, and nor do I claim any equivalence between vegan and anti-vegan hostility. But these examples are indicative of a direction the debate *could* head. It is my suggestion that a non-vegan, although animal-rights-respecting, food system offers a vision of the future palatable—more, *desirable*—to everyone. It satisfies animal advocates as it respects animal rights. And it satisfies critics of veganism because it allows them access to non-vegan foods, work with animals, and so forth.

Later in the chapter, I come to why this possibility is desirable for animal advocates, but let us note now that it is surely prima facie desirable for *critics* of veganism. It is desirable relative to a vegan future, as it allows them access to the things veganism closes off. And it is desirable relative to the current situation as it allows them to make good on claims to regret animal suffering and death. Take, for example, Matthew Evans, a food journalist, chef, farmer, and apologist for meat-eating. He sincerely believes that meat has culinary and cultural value; that working with animals is meaningful and important; that veganism may be bad for animals, the hungry, and others—some of the things explored in this chapter.

However, Evans supports 'minimising harm [and] rearing fewer animals for the same gastronomic and nutritional benefit' (2019, 209). He also regrets animal death: 'There is nothing, I repeat *nothing*, nice about seeing a warm-blooded animal take its last breath' (2019, 97, emphasis Evans's).

Although Evans is no animal advocate, a world in which the goods associated with animal agriculture remain *but* animals do not face suffering and slaughter would (surely?) be his arguments' natural conclusion.

In offering a principled separation of animal rights and veganism, I offer a vision of animal rights more attractive to a sympathetic audience than an old abolitionist conception. Here, I echo Donaldson and Kymlicka. They observe that abolitionists' 'extinctionist' commitment to companion animals turns away potential allies drawn to animal rights because of feelings towards companions. Arguments about the end of cats and dogs are off-putting—and hand ammunition to abuse apologists, who disingenuously paint animal advocates as animals' *real* foes (Donaldson and Kymlicka 2011, 9–10). Similarly, *anyone* drawn to animal rights may be put off by an insistence (or hidden conclusion) that animal rights means universal veganism—and, thus, the end of foods, practices, work, sights, and more that people find valuable.

If advocates can show potential allies that the goal of animal rights is *not* the elimination of these valued things (and that animal rights need not mean interhuman injustices), these potential allies will be readier to 'sign on'. If Donaldson and Kymlicka's zoopolitics can attract converts by telling them they can keep their beloved companion dogs, imagine the converts we could attract to a zoopolitics allowing people to keep their beloved companion dogs *and* their beloved hot dogs.

I do not aim to convince readers to adopt my approach because doing so has pragmatic advantages. I rest my case on *principles*. Thus, although these pragmatic advantages are real, I now leave them aside, and turn instead to the core question of this chapter: what *principled* reasons do we have for exploring the possibility of a non-vegan food system? First, I look to liberalism. Second, I look to animal rights.

Liberal problems with veganism: Disagreement

Humans each have their own (to invoke Rawls) *conception of the good*. For current purposes, we can think of this as an understanding of what a good life looks like—of what would give our life meaning. Religious commitments, conceptions about the place of the family, beliefs about the importance of hard work, convictions about the relative value of different pursuits, and more are what we use to order our lives, and, together, they constitute our conception of the good.

Generally, liberal political orders endorse the centrality of conceptions of the good without asserting that any *particular* conception is correct. (More on this in Chapter 7.) So, it is one thing to endorse state support of higher education because the enriching experiences of learning help people recognize and seek out their own good. It is another to endorse state support of higher education because the good life is one of learning. Whatever the strengths of the arguments, one relies on humans' diverse conceptions of the good, while the other relies upon a particular conception of the good. The former is recognizably liberal. The latter is not.

There are reasons to think that a vegan society will contain limited opportunities for many humans to realize their particular conceptions of the good. I am not saying that people might be *unhappy* with or *inconvenienced* by veganism. I am saying that people (sincerely believe that they) will not be able to lead a good life (whatever that means to them) if the state imposes veganism.

Now, *even if they are right*, this does not prove the imposition unjust. It may be that some conceptions of the good are unrealizable in a just state. A would-be warlord's good life might involve land grabs, skirmish, looting, and slavery—but these are unjust. In principle, conceptions of the good closed off by veganism may be relevantly like the warlord's conception. But even were that the case, it would be *regrettable* that the demands of justice closed off the realization of an individual's good. 'Alas', the liberal might say, '*if only* we could have a situation in which the warlord was free to pursue his conception of the good *and* we realized justice'.

I return to these thoughts at the end of the chapter. For now, it is worth exploring the idea that veganism *does* shut off the realization of certain conceptions of the good. And not just for a few eccentrics, but for many of us. Four examples of elements common among conceptions of the good follow. I stress that these are *examples*, and do not claim that they are exhaustive. That they are *common* is important. Philosophers are fond of examples of individuals who find meaning in the most bizarre things. However, the four issues explored are not things important only to oddballs. Instead, they are questions central to the lives of many real people.

Disagreement, part 1: Food practices

Almost all of us value particular food practices. The value can come from elements beyond the obvious—beyond the biological importance of eating and the aesthetic value of taste (Barnhill et al. 2014). Food practices might be important parts of relationships with others, including others who are no longer with us: barbecues with Dad, ice creams with Nana, a favourite restaurant with old friends, a favourite meal with a lover—all may be important to our interpersonal relationships, and these relationships may be identity-defining and meaning-making.

These barbecues, ice creams, restaurants, and meals may not be vegan-friendly. And even if the ice cream could be a sorbet and the restaurant can serve falafel, this change may influence the relationship. The experience just might not be *the same*—and the change could harm the relationship (Emmerman 2019).

Meanwhile, food can be integral to valued occasions. To draw on some examples from my own cultural background, birthdays are associated with cakes; Hallowe'en with sweets; Easter with lamb (and chocolate); Christmas with turkey (and more); football with pies; midsummer with barbecues. Again, these occasions may be valuable in-and-of-themselves, may be important for interpersonal relationships, and may form a part of our wider identity. Tweaking the food practices—nut roast rather than turkey, say—could adversely impact the whole practice. Christmas without turkey just might not be Christmas anymore.

Vegans might contend that *they* could switch from turkey to nut roast without loss of value—but that is not to say that others will have the same experience (cf. Ciocchetti 2012). It is not for me to say what is *really* valuable for others. Individuals can and do have importantly different conceptions of the good, even if they are from similar backgrounds.

Food practices are important, too, for our cultural identities. Deli meats and cheeses may be important for Italians; sausages and stews for Germans; dried fish for Norwegians (cf. Nguyen and Platow 2021). People with strong sub-national identities, too, can express this through food: Morecambe Bay shrimp and cockles, hotpot, and butter pie may be important for Lancastrians. Certain foods might also have identity-defining class associations—the strong working-class (and northern English) association of the British chain Greggs is part of the reason that the already-mentioned vegan sausage roll caused a furore.

Professional or lifestyle identities can be strongly associated with non-veganism. I discuss the importance of work below, but, for now, consider the men for whom meat-eating is part of their masculine identity. Whatever readers might think of these men's attitudes,[4] their gender identity may be an important part of who they are, and losing their ability to express this may

---

[4] There has been considerable scholarship critical of the link between meat and masculinity. I discuss some of this in later chapters. My point here is simply that masculinity need not be all bad, and if meat is associated with masculinity, that could be a mark in meat's favour—at least for those who value their masculinity.

be bad for them, impacting their ability to pursue their own idea of the good life. For them, an effeminate life may be a poor one (for a man), and a life without meat may be effeminate.

For liberal philosophers, religious identities are particularly important. Religious food practices are frequently non-vegan: chicken soup, for example, can lead to tension for Jewish vegans, and ghee has both ritual and cultural significance for many Hindus. Animal advocates might claim that religious practices do not *require* these products—'nobody is bound to eat meat' (Barry 2000, 45). There are two problems with this. First, there is room for theological debate. We can read Augustine and Maimonides, central pillars of Christian and Jewish thought respectively, as saying that killing animals for meat is morally necessary (Lagerlund 2018, 764). Second, however, this is not the point. The important matter is not that individuals are (or feel) obliged to follow a non-vegan diet. The point is that people value these food experiences as part of something independently important, such as their religious or cultural identity—things that are part of their conception of the good.

Foods and food practices we associate with particular people and particular activities are, basically by definition, occasional. Practices that we tie up with our religious, cultural, or other identity, on the other hand, are 'everyday'. Everyday food practices can also have importance to us in the most secular sense. They provide stability and assurance; they mark points in the day or week; they get us ready to work, relax, begin, sleep, or what-have-you. If, without fail, I have a latte before starting work, this could take on its own value, and become close to a necessity for me to work. And my work will—if all's well—have value. It will be bad for me if my latte is not available, or if I must replace it with dairy-free alternative. This is, on the face of it, regrettable. It impacts my efforts to live well.

Disagreement, part 2: Taste

Animal ethicists traditionally downplay the value of taste. Tom Regan observes the flippancy with which Peter Singer dismisses 'the pleasures of taste' as 'trivial'. On the contrary, says Regan, 'Many people, including quite thoughtful persons, do not regard the situation in this way. Many go to a great deal of trouble to prepare tasty food or to find "the best restaurants" where such food is prepared' (2004b, 220).

Critics of veganism sometimes defend meat-eating with appeals to the value of eating good food (e.g. Lomasky 2013), painting the gourmet and the vegan as natural enemies. More than one celebrity chef has expressed open disdain for veganism (Wright 2015, chapter 1). Susan Wolf (2018), a philosopher and foodie, is more measured. She allows that, compared to animal harms, the ethics of foodie-ism seems trivial. However, she offers a fair defence of the foodie lifestyle and its virtues.

The foodie conception of the good is welcome in a liberal state, yet the foodie surely is not a natural vegan. Veganism cuts off foodstuffs, and (almost?) entire cuisines. (I have cookbooks dedicated to 'veganizing' a range of cuisines, but I do not have *Vegan Inuit Meals*.) Veganism makes it difficult, even impossible, to access certain flavours, textures, and forms of cooking. It means an end to many foodie favourites: Wagyu beef, foie gras, dairy cheeses, oysters, caviar, *fugu*, blue-fin tuna, lobster, Iberian ham, and more. To Singer, these losses are trivial. To the committed gourmet, they are catastrophic.

One need not be a foodie to recognize the importance of taste. Jean Kazez (2018) asks us to imagine a nutrient-rich but uninspiring 'perfectly ethical food' (PEF).[5] Would our life lose something if we were to switch to eating only PEF? Would our life be less meaningful? Perhaps taste alone is a reason not to eat PEF—even if that meant we valued taste more highly than some ethical demands, switching to a food that is *not* perfectly ethical. If we should still favour PEF, imagine a variant in which PEF is not merely uninspiring, but only just palatable. Or another version in which it is foul. Kazez's point is that taste carries *some* weight. The goods of taste *could* outweigh ethical demands upon us. Or—to put it in the language of this chapter—it is perfectly reasonable that, for many (most, all) of us, good-tasting food has a part to play in the good life. In (almost) all of us, there is a bit of a foodie. All of us have reason to lament the loss of diverse and intense food experiences.

And foodies (of the Wolf *or* Kazez variety) have support from philosophical aesthetics. While philosophers have denigrated taste compared to other senses (and culinary arts compared to other arts), philosophers of food, philosophers of art, and aestheticians increasingly accept food as art and the aesthetic value of taste. Even if some readers might see finding life-defining value in food experiences odd—the foodie might sound more like an eccentric than (say) someone committed deeply to their church—we can all recognize finding life-defining value in art and aesthetic experience as reasonable.

---

[5] Many vegans (and non-vegans) eat something resembling PEF: meal replacers with names like Huel ('Human fUEL'), Soylent (which is not people, unlike its namesake), and—a word familiar to fantasy buffs—Mana.

Disagreement, part 3: Work

For many of us, our work is our life. Even if work is just something we do, work is a 'central arena of self-making, self-understanding, and self-development' (Clark 2015, 61). Some work is good, some is bad. Maybe we can measure this objectively—good work might require us to exercise our capacities, have a degree of control, etc.—or it might be up to the worker. Either way, a vegan state is one in which much good work is lost. (We can distinguish good work from *righteous* work, and allow that there is good work that is unjust (Clark 2015).) And removing these sources of good work stymies people's pursuit of the good.

Good work relying on non-veganism is not hard to imagine. The crofter working with small herds; the craftsperson building dry stone walls for shepherds; the patisserie chef cooking with eggs, milk, and gelatine; the small-business owner selling fine cheeses; the academic involved in agricultural education; the journalist writing on 'rural affairs'; the investor speculating on pork futures; the hotelier serving award-winning bacon—all might have good (albeit differing) jobs. All might fairly complain that they lose good work in the vegan state, and that universal veganism scuppers their chance of living a good life.

Jocelyne Porcher is one academic critical of veganism because of work. For her, husbandry—small-scale animal farming, an 'idyllic' vision of what agriculture could be—is worth defending because of the value it brings to farmers. She can barely express this value in words:

> How difficult it is to make others feel what it is to live and work with animals when it is not money but happiness that ties you to animals and to the world? How can these emotions, these fragrances, these tastes, this physical contact and these sounds be expressed? It is a whole universe of sensations experienced daily. How can you describe the way a ewe looks at her newborn lamb, and at you, who are there vigorously rubbing the lamb with a handful of straw so that it will not be cold and will want to suckle? How can you describe the scent of the fleece of a ewe that wafts behind the herd as they walk, and mixes with the misty morning air, a fragrance that is heady and sweet, that returns to me as I write and that I believe I can detect around me in my office now? How can you describe the sense of space that shepherds have as they traverse the mountains with their sheep: the solitude, the fear and the perfect happiness? What can we say about the ties with animals that have a thousand faces, a thousand forms and a thousand places?
>
> (Porcher 2017, 122)

A world without this work is, for Porcher, impoverished. Indeed, a world without it, she says, renounces life itself (2017, 122). I interpret this as a claim about her conception of the good life. Whatever the merits of her arguments,[6] she has a conception of the good that the vegan world challenges. For her, work with animals is central—and this is not, she thinks, a personal idiosyncrasy, but a part of the character of pastoralists (smallholders, country folk …) generally.

To jump ahead, I do not pretend that the world I envision will contain *all* the non-vegan work this world currently contains. Most starkly, I do not believe that the world I envisage would contain slaughter-work. This, however, is no great loss. Slaughter-work is some of the worst available, with physical danger from whirring blades and kicking animals, and emotional and mental strain from repeated killing (see, e.g., Joy 2009, chapter 4). Few enjoy the work. In the words of Evans—who defends slaughter—anyone who does 'probably shouldn't be anywhere near a kill' (2019, 97). Even at its best, slaughter is grim work (Evans 2019, chapters 5 and 10). Perhaps the non-vegan but animal-rights-respecting state is able to have the best of both worlds; perhaps it can retain the good work of the non-vegan world, but lose the parts of the food system that are regrettable even to the advocate of meat eating. Maybe a non-vegan zoopolis can have both *good* and *righteous* work; it can be a state with *humane jobs* (Coulter 2016, 2017).

Disagreement, part 4: Land

For me, 'countryside' evokes images reminiscent of England's Lake District. Rolling, grassy hills; sharp outcrops of rock; babbling brooks leading to placid lakes; sheep. As a child, I did not recognize that this vision, as well as being far from universal, is of an artificial environment. It is an environment shaped by humans, and, crucially, by pastoral farming. These hills and mountains are the shape they are because of centuries (millennia) of human use.

While some of this use, from an animal-rights perspective, is relatively innocuous—hiking, say—much involves violations of animals' rights. Farmers build walls to contain animals. Engineers redirect water to create natural barriers or drain bogs. Landowners shape and stock lakes for anglers. Sheep

---

[6] For a compelling rebuttal, which brings her vision more in line with my own, see Delon 2020.

eat plants, or do not, while shepherds remove trees and underbrush to create sheep-friendly spaces. Sheep, shepherds, dogs, horses, traps, quadbikes, and four-by-fours beat paths. Without farming (and fishing, and the rest), the landscape would look very different.

So what, the vegan might ask? First, people *want* to see animals in farmland (Rust et al. 2021). They do not want to see abattoirs, factory farms, or manure lagoons, but they *do* want to see domesticated animals. A vegan world is one without these animals, so a little less bright. (I return to a related thought later.) Philosophical critiques of veganism rarely draw upon this idea, although there are exceptions. For example, Chris Belshaw—a philosopher whose 'countryside' is not unlike my own—reflects upon the importance of animal agriculture by saying that

> Though enjoying meat isn't a part of the good life [foodies beg to differ!], enjoying the beauties of nature arguably is. Think about the Scottish Highlands, with mountains, heather, grouse, and deer. The image here is not of untrammeled [sic] nature but of a nineteenth-century construct, close on the heels of [mass evictions in the eighteenth and nineteenth centuries], [Walter] Scott's novels, and [Queen] Victoria's enthusiasm. Whatever its provenance, many think it important that it be sustained. Curiously, some people will pay to shoot grouse and deer. Less curiously, others will pay to eat them. Insofar as we support these practices we help preserve this landscape. Without them economic pressures for a very different use of the land are harder to resist.
>
> (Belshaw 2015, 20)

Here, Belshaw places aesthetic appreciation of 'nature' higher than aesthetic appreciation of meat. And, as he rightly points out, the aesthetic appreciation of (a particular kind of) countryside is possible only because of industries violating animals' rights.[7]

How might the vegan respond? A vegan society could allow for aesthetic appreciation of the countryside; it would offer an *alternative* beautiful landscape. Or a vegan society could maintain the *present* landscape without rights-violating industries. Belshaw, I think, would not be impressed. The former is like the vegan assuring the venison-lover that they can enjoy tofu on a vegan diet. But the gourmand does not *want* tofu—they want venison.

---

[7] To jump ahead, I envision a food system with venison sausages, seared trout, roast grouse, and the rest. But does my vision allow shooting or angling? In the spirit of this book's argument, I support developing rights-respecting alternatives to these pastimes. For example, 'hookless fly fishing' and 'fish watching' seek to replicate the positive aspects of angling without fish suffering and death. Perhaps, in the zoopolis, there will be anglers, they just will not hurt fish.

And while the vegan zoopolis *could* fund managers to maintain Belshaw's Highlands or my Lakelands, we can forgive Belshaw for not holding his breath.

Is this all just a preference about views from a train window? How important are these claims? Well, let us not be too quick to dismiss the value of aesthetic experience, which can be critical to the good life. The Highlands and the Lakelands have their share of artists singing (literally and metaphorically) of their sublimity. However, there are at least two reasons to think that there is more than aesthetic experience at stake. The first relates to an image of the importance of (human relationships with) *nature*. The second concerns the *nation*.

An understanding of (our relationship to) land is core to many approaches to environmental ethics. It is no coincidence that Aldo Leopold's ethical framework is the *land ethic*. According to this approach—and associated conception of the good—'a thing is right when it tends to preserve the integrity, stability, and beauty of the biotic community. It is wrong when it tends otherwise' (Leopold 1949, 224–225). Without delving too far into the land ethic, it is easy to imagine how it could support meat-eating, especially if the 'biotic community' looks like my Lakelands or Belshaw's Highlands. Twenty-first-century environmental ethicists have not shied away from using the land ethic to defend meat-eating (e.g. Callicott 2015; List 2018)—although it is perhaps easier to defend hunting than pastoralism.

And an image of the land can be as central to an understanding of *nation* as to an understanding of *nature*. (*Nation* and *nature* share the Latin root *nascor*, meaning 'origin'—think of *natal*. What is our origin if not the land?) For nationalists, the good life is a life lived with her people, in the way her people live, *in the place* her people live. A change in the land, meaning it is no longer recognizably *her* land, is thus a great evil.

And not just for the nationalist. There is a reason there is something wretched about 'fourth world' peoples. Not only do they lack access to *any* land, but they lack access to *their* land. Constructed or not, images of the Highlands might be deeply important for the Scottish nationalist; images of the Lakelands might be deeply important for the English nationalist. Both might be important for the British nationalist, or Celtic nationalist, or similar. For nationalism's critics, these feelings might be irrelevant, but liberals cannot dismiss nationalist conceptions of the good; a nationalistic conception need not be incompatible with justice (as the warlord's is).

The reasonableness of the value individuals place on aesthetic appreciation of the land, on connection to nature, and on connection to the land of *their people*, show the importance of images of the land. That the possibility of the vegan state threatens these conceptions is something that the liberal must take seriously—whatever her own view.

Disagreement, parts 1–4

These examples—food practices, taste experience, good work, the land—show that veganism poses a threat to many reasonable conceptions of the good. It is not just eccentrics who hold these conceptions of the good—ordinary people do, too. Indeed, we might think that someone who does *not* consider food and work practices or aesthetic appreciation part of the good life is the eccentric—although my argument does not rest on this stronger claim.

Again, these are just examples. But, together, they show why we should take seriously the possibility of a non-vegan food system—even if we believe in animal rights.

Liberal problems with veganism: Food justice

As I noted in the book's introduction, vegans and animal activists can have a testy relationship with food justice activists. In some cases, food justice activists do not necessarily object to veganism, but do object to vegan activism. Animal activists might be ignorant of food justice concerns that do not directly involve veganism (Mares and Peña 2011). And misdirected activism can perpetuate food injustice, as when activists single out food practices of already-marginalized communities (see Kim 2015), or when activists harshly criticize people facing significant barriers to veganism for their food 'choices'. However, the fact that activists face (legitimate) criticism for ill-thought-through activities does not mean that their aims are misguided.

There are, however, more principled food-justice objections to plant-based food systems. Vegans might fairly be accused of focusing on veganism for 'normal' (able-bodied, white, male …) consumers, and overlooking the barriers to veganism for others (George 1994, 2000). Women (especially when pregnant or breastfeeding), children, the disabled, athletes, and non-white people may all have dietary needs making veganism tricky. In turn, the imposition of veganism may disadvantage these groups.

Veganism may also scupper attempts to deal with existing food injustices. Take world hunger as a stark example. According to The Food and

Agricultural Organization of the United Nations, of the 7,633 million people in the world in 2018, 2,014 million of them were at least 'moderately food insecure',[8] with 704 million of this group 'severely food insecure'[9] (FAO 2019, 20). There are at least two reasons to worry that veganism could exacerbate this. The first is that meat is a nutrient-dense food useful for the malnourished. Evans uses the example of vitamin A:

> Globally, 155 million children are stunted and 52 million children suffer wasting, simply as a result of diet, and 88 per cent of countries suffer the burden of several forms of malnutrition. A third of all children in the world are deficient in vitamin A. Of the top 20 dietary sources of vitamin A, there's not a single vegan option. Liver, particularly from beef cattle, is one of the best sources. Our bodies can covert nutrients from plants into vitamin A, but you need to have access to those plants, to cook them, to eat enough of them (because it takes 12 times more carotenoid in a plant to get the same amount of vitamin A from an animal source), and then also consume those vegetables with fat to ensure the nutrients are absorbed into the body. The quickest, simplest way for impoverished people to get enough vitamin A is to give them animal products. Veganism isn't even on the agenda when basic health is at stake.
>
> (Evans 2019, 195–196)

Evans identifies another problem: large areas of land are not suitable for growing crops, even while they *are* suitable for pastoral farming (2019, 184–187). Dictating to the most impoverished that they cannot farm the food to which their land is most suited threatens to worsen their malnourishment, in turn making the community more dependent on food aid or trade, and thus eroding their food sovereignty. (It also separates them from established ways of life, running into the problems discussed earlier.) Imposing veganism thus runs the risk of intensifying or creating serious food injustices. This is not a matter of people not being able to eat what they like; it is a matter of life and death.

Justice questions also arise when state action removes, without replacing, good work. As noted, there is much good work that universal veganism removes. What is more, this good work might be disproportionately held

---

[8] 'The level of severity of food insecurity, based on the Food Insecurity Experience Scale, at which people face uncertainties about their ability to obtain food and have been forced to reduce, at times during the year, the quality and/or quantity of food they consume due to lack of money or other resources. It thus refers to a lack of consistent access to food, which diminishes dietary quality, disrupts normal eating patterns, and can have negative consequences for nutrition, health and well-being' (FAO 2019, 188).

[9] 'The level of severity of food insecurity at which people have likely run out of food, experienced hunger and, at the most extreme, gone for days without eating, putting their health and well-being at grave risk, based on the Food Insecurity Experience Scale' (FAO 2019, 189).

by disadvantaged groups. Take Porcher's husbandry. People in the countryside overwhelmingly hold this work. People living in rural communities might fairly complain that the imposition of veganism unfairly disadvantages them—potentially an already disadvantaged group—relative to urbanites. If pastoralists were left unemployed, the injustice would be clear. But even if they were able to take up other, less good jobs—menial work producing canned vegan soups, say—the injustice would remain. Indeed, being left out of work or having only bad work may be an injustice itself. Again, perhaps especially so if members of disadvantaged groups are disproportionately affected.

If veganism risks exacerbating hunger and food insecurity, as well as eroding food sovereignty, the food practices of minority groups, and good work with food, it is no wonder that animal-rights activists and food-justice activists have not always seen eye-to-eye. What does this mean for us? There are five possibilities.

First, we could reject one of either animal rights or food justice. This is not compelling. In both cases, the injustices are real.

Second, perhaps the challenge presented by food justice is not as dire as it sounds. For instance, it is sometimes argued that the inefficiencies of animal agriculture mean that plant-based food systems are a promising antidote to food insecurity (Shepon et al. 2018). But it would be naive of animal activists to pretend that there will be no conflicts in a shift towards veganism.

Third, animal rights theories could adapt, taking food justice seriously, and ensuring that a shift to veganism has a minimal negative impact on humans. For example, perhaps animal-rights frameworks should be complemented by the already-mentioned 'humane jobs' approach, ensuring that vegan jobs are *good* jobs, and ensuring that a vegan economy is not one in which people are left with only bad work (Coulter and Milburn 2022). This approach has merit. Surely, the best case for a vegan food system would be one that adapted to avoid food injustices.

But even in the best vegan system, food justice concerns may remain. A fourth option sees a careful balancing of injustices, allowing that sometimes (regrettably) animals lose out, and sometimes (regrettably) humans lose out. Analogously, to use a well-worn example, a starving person can steal from a wealthy person to feed her family. The starving person ignores the wealthy person's property rights (an apparent injustice), and a family is starving (another apparent injustice). In this case, the injustice of starvation likely outweighs the injustice faced by the wealthy person.

Some rights jargon is helpful. In normal circumstances, stealing would *violate* property rights. In this case, however, the owner's property rights are merely *infringed*. This does not mean that her rights vanish. The poor person must favour a non-rights-infringing third option if there is one. (For example, she cannot steal if charity is an option.) It also means that the poor person may owe the wealthy person an apology or recompense, even though the theft was permissible. And so on. These things would not be the case if the rich person's rights had vanished.

So perhaps members of disadvantaged human groups might continue (say) hunting to avoid the imposition of stark food injustices. While such hunting would normally *violate* animals' rights, here it only *infringes* them. Such conflicts are tragic; either way, someone has their rights trampled, even if that trampling is an infringement only, and thus not unjust. This works as a last resort in the messy real world, but, as political actors seeking just possible worlds, we can aim for something better.

Readers may have guessed option five. The existence of these possible food injustices gives animal rights theorists principled reasons to examine the possibility of a non-vegan, but animal-respecting, food system. More, they give the zoopolis a reason to *favour* a non-vegan food system, *if* it is likely to contain fewer food injustices than vegan food systems.

Animal-rights problems with veganism: Harm to animals

That some humans have reasons to object to a vegan world is not surprising. To suggest that a vegan world could be contrary to the rights of *animals*, on the other hand, sounds odd—but perhaps it could be.

Anyone who has spent time around arable agriculture knows that farming plants leads to harm to animals. Farmers clear woodland, drain bogs, and dig up the earth. These activities destroy animals' homes and foraging grounds. Traps, poisons, and guns kill animals threatening crop production. And agricultural machinery kills and maims animals—as any large machinery operating in animals' spaces will.

These observations lead to crude arguments that animal advocates should prefer pastoral agriculture to arable agriculture, or that the optimum diet to prevent harm to animals contains meat. Widely shared articles with titles like 'Ordering the vegetarian meal? There's more animal blood on your hands' (Archer 2011a) lead to people sincerely believing bizarre claims about relative numbers of animals killed in arable and pastoral systems.

We can easily rebuff critics who present this as a 'Gotcha' against veganism. Most animal agriculture relies on intensively grown crops to feed animals, leaving pastoral agriculture *doubly* harmful. But a more sophisticated set of 'new omnivore' arguments warrants more careful consideration (Milburn and Bobier 2022). Some new omnivores point to particular forms of pastoral farming, such as farming cows on pasture, as less harmful than plant-based food production (Davis 2003; Archer 2011b). Such arguments cannot be allowed to have the last word, as they rely on questionable calculations of numbers killed (Matheny 2003; Fischer and Lamey 2018) or over-simple ethical principles (Lamey 2019, chapter 4).

This does not mean that field deaths are irrelevant for discussions about the ethics of eating. Donald Bruckner (2015; cf. Abbate 2019; Milburn and Fischer 2020) argues that eating 'roadkill' does not violate animals' rights, and, given that we harm animals when farming plants, vegans should eat roadkill *rather than* plants. Consequently, by the lights of the vegan's own arguments, veganism is immoral. Similarly, Christopher Bobier (2019) argues that the vegan's own argument against supporting industrial animal agriculture generalizes to an argument against industrial *vegetable* agriculture. Others nod towards the harm of arable agriculture to justify eating products of cellular agriculture (Milburn 2018), products of hunting (Calhoone 2009), eggs (Fischer and Milburn 2019), invertebrates (Fischer 2016a)—and so on.

The relevance of these moral arguments to this political project is worth spelling out. There are potential elements of a vegan food system—intensive arable agriculture, for example—that are harmful to animals. Some of these harms will be rights-violating. The poisoning of rodents and the shooting of birds will often[10] violate rights. Farmers destroying animals' homes may also violate rights, although this is complicated.[11] Even when these harms to animals are not straightforward rights violations—unintended harms to animals hit by harvesting machinery, for instance, may not constitute rights-violating violence (cf. Lamey 2019, chapter 4)—there may be issues of justice at stake. For example, animals may have rights to have their good counted as part of the public good, and to have their voices represented in political decision-making procedures (Cochrane 2018). States would not see the deaths of field animals as inconsequential if they respected *these* rights.

This takes us into complicated questions about what rights animals have—and the question of separating those duties we owe to animals as a matter of *justice* from those that are a matter of (mere) morality. But let us be clear on what the harms to animals in arable agriculture mean for the present enquiry. Much arable farming involves harm to rights-bearing animals, and these

---

[10] Surely I can shoot animals if that is the only way I can protect the crops I need to live?
[11] See Hadley 2015; Cooke 2017a; Milburn 2017a; Bradshaw 2020; Kianpour 2020.

harms may constitute (or result from) injustices. Consequently, an unreserved endorsement of arable food systems is problematic. We must work to identify which arable systems involve (rights-violating) harms, and it may be incumbent upon the zoopolis to support only *some* forms of arable agriculture.[12] Conversations about reducing harm to animals in arable agriculture are worth having—and this is as much a socio-political issue as a moral one—but here is not the place for it (cf. Milburn 2022b, chapter 5).

Parallel to a conversation about animal-friendly arable agriculture, however, is the possibility of non-vegan food systems that are nonetheless animal-friendly—perhaps *more* animal friendly than some (many? *All*?) arable food systems. Violations of animals' rights in arable systems give animal ethicists principled reasons to examine the possibility of a non-vegan, but respectful, food system. More, they give the zoopolis a reason to *favour* a non-vegan food system, *if* it could involve fewer rights-violations than a vegan system.

Animal rights problems with veganism: Extinction

The abolitionist vegan future of the old conceptions of animal rights features few animals. There is something paradoxical (although not necessarily wrong) about respecting animals' rights by eliminating them. But it is not merely paradoxical. It is deeply unappealing. A future in which domestic animals no longer exist, in which humans interact with animals (if at all) only through fleeting encounters with wild animals, hardly sounds like a future to strive for. Indeed, for some readers, it will be dystopic.

When meat apologists use the existence of animals as a justification for meat eating (see, e.g., Zangwill 2021), the conversation turns to interesting (although difficult) questions about the merits of existence vs. non-existence, asymmetries between obligations to create happy beings and obligations *not* to create unhappy beings, and so on. I will touch on these debates in Chapter 6, but I will say no more about them here. For now, I base my argument on the intuitive appeal of a future in which humans and animals live together.

It is possible that humans in the vegan zoopolis interact with *some* animals. For example, maybe a vegan zoopolis requires us to (almost) entirely cease relationships with farmed animals, but retain relationships with companion

---

[12] Activists already push in this direction, but their concerns are selective. For instance, many vegans are concerned about orangutans' habitats being destroyed by palm-oil production, but are less concerned about birds sucked into olive-harvesting machines. Hence, they object to palm oil, but not olive oil.

animals. Perhaps this is where Donaldson and Kymlicka's vision of the zoopolis leaves us. Despite talk of human relationships with domesticated animals, they are seemingly talking mostly about dogs, cats, and similar. Recall Garner's comments I quoted in this book's introduction; although I defended Donaldson and Kymlicka for leaving *some* room for farmed animals, they do not leave *much* room. People in the zoopolis would have little reason to keep cows and pigs if there was no market for their products. Garner is half right; Donaldson and Kymlicka's zoopolitics retains a *flavour* of abolitionism.

A society with companion animals but no animal protein may be questionable—what will the companions eat (Milburn 2015, 2022b, chapters 2–3)? But my point here is that (historically) farmed animals face near-extinction on Donaldson and Kymlicka's picture. Yes, as Donaldson and Kymlicka note, 'there will always be people who want to have cow companions' (2011, 139), and have the space, time, now-how, and resources to support these animals. And there will be the occasional town welcoming sheep to keep grass short (Donaldson and Kymlicka 2011, 136). But ultimately, Donaldson and Kymlicka's zoopolis—although not *strictly* extinctionist—means near extinction for the animals who have been (arguably) both the most important in the emergence of a de facto mixed-species community *and* those who have been most exploited by humans. Intuitively, more desirable than this would be a society in which this group had a home—*that* is what we would *expect* rights to mean.

The value of this observation goes beyond intuition. It is more in line with Donaldson and Kymlicka's arguments to avoid quasi-extinctionist conclusions. They compare animals in human society to humans descended from those who have been pulled into a given society against their will and treated as inferior. Acknowledging that the society should have recognized these individuals from the offset does not mean that the society should force these humans back to where they came from (they now have no home but in the new society) or that they should be exterminated (the idea is abhorrent). It means the society should welcome them as co-members (Donaldson and Kymlicka 2011, 79). This is a core argument for Donaldson and Kymlicka's claim that domesticated animals are co-community members—indeed, for one commentator, it is their *only* argument for the conclusion (Sachs 2019, 94–95).

As an argument for welcoming domesticated animals into our society, I think this is effective. However, there is an important consequence to this argument. We should try to find these newly recognized members of our society *a place*. 'You are one of us now' is hollow if not accompanied by 'and here's a place for you'. Even a vegan society with dogs and cats seems to say, to many animals, 'You are one of us now, but we have no place for you'. In Chapters 5 and 6, I introduce the idea of formerly farmed animals as workers. This possibility gives these animals—arguably the most exploited, and the ones that the vegan state would see vanish—a space in which to live and flourish in a non-vegan zoopolis. For the advocate of animal rights, that must be an advantage.

Conclusion: Tying threads together

Let us bring together the different parts of the argument. There are pragmatic concerns we could raise about veganism. But I am interested in *principled* worries about the vegan state. I have canvassed four. The vegan state, I have argued, (1) cuts off a range of sources of a valuable life; (2) introduces or exacerbates food-justice concerns; (3) continues to harm animals through arable agriculture; and (4) counter-intuitively, will lead to human/animal separation.

That a vegan state may cut off sources of value—some will be *unable* to live (their vision of) a good life—is not proof that the vegan state is not what justice demands. Recall the earlier example of the warlord. Perhaps some individuals are unable to live according to their conceptions of the good in a just state, because their good necessarily entails injustice. But we should regret the fact that laws (would) cut some off from a meaningful life. It gives us a reason to explore whether a non-vegan state may be compatible with the demands of animal rights. Scholars of animal rights should not be satisfied with the prospect of a vegan state. While this is *one* state apparently compatible with animal rights, they have good reason to seek another.

We can go further. Assuming we can envisage a rights-respecting non-vegan system, the zoopolis has good reason to *permit* it. As noted in the introduction, we should not endorse larger states than necessary (cf. Machan 2004, 23), and nor should we readily forbid actions that do not violate rights. By the invisible hand of the market—drawn by the many conceptions of the good calling for non-vegan foods—permitting non-vegan foods in the zoopolis would ultimately lead to a non-vegan food system. (I will say a little more about *permission* in Chapter 7.)

And we can go yet further. The zoopolis must be ready—in principle—to *endorse* and *support* non-vegan food systems. Why? This chapter has canvassed at least two potential sources of injustice in vegan food systems: food justice concerns and harms to animals in arable agriculture. It is plausible that the non-vegan system could overcome these injustices. That alone is enough for the zoopolis to support the rights-respecting non-vegan system. But it is not *only* justice that is relevant, here. Perhaps liberal states can legitimately support food systems if doing so will help citizens live good lives. (Again, more on this in Chapter 7.)

Assessing the full extent to which animal-rights-respecting non-vegan food systems and animal-rights-respecting vegan food systems might overcome the challenges explored would be a colossal undertaking. I cannot achieve it in this book. I certainly offer *indications* of why a non-vegan system may overcome these challenges better than a vegan system, but allow that, ultimately, a vegan system (likely: a radically reimagined vegan system) may be able to overcome these challenges better than the best conceivable non-vegan system. I would consider that surprising, however. After all, a non-vegan system can incorporate all that a vegan system can—and then some. Ultimately, I suggest, it is unlikely that the all-things-considered best food system happens to be vegan.

My inability to prove decisively that a non-vegan food system would be better leads me to the final element of this chapter's argument. It is an argument relying, first, on intuition. I have argued that there is something odd about the (near-)extinctionist conclusion of the vegan position. This does not prove that a vegan state is not what animal rights demands. But it does give us *some* reason to endorse a non-vegan state over a vegan state. If we are in a position in which both a vegan state and a non-vegan state appear to be (somewhat) consistent with the demands of justice, we should go with the one that seems a little less *anti-animal*. This is also the conclusion reached by recognizing animals as members of our society. We should be ready to give them *a place*.

The conclusions about permitting and supporting a non-vegan food system rest upon the claim that an animal-rights-respecting yet non-vegan system is possible. Much of this book will be given over to arguing that it is. I address possible elements of such a system in turn, showing which are compatible with animal rights. By the end of the book, there will be a rich (although perhaps not full) picture of a possible system—a system that, if the arguments of this chapter are compelling, the zoopolis could (should!) endorse and even support, although it is far from vegan.

2
Bugs and bivalves

Not all animals have rights. At least, it would be strange to hold that membership in the kingdom Animalia is necessary and sufficient for being a rights-bearer. This would simply take one arbitrary biological grouping as the foundation of rights—membership of the species *Homo sapiens*—and replace it with another. Instead, claim animal rights theorists, *sentience* is a necessary and sufficient condition for the possession of rights. People holding such a view are *sentiocentrists* or—my preferred term—*sentientists*. Although the phrase does not have the same rhetorical force, they are advocates of *sentient rights*, not *animal rights* (Cochrane 2013).

Veganism, as I understand it (Dickstein and Dutkiewicz 2021), does not distinguish between sentience and non-sentience. Vegans abstain from *animal* products, not merely products from *sentient* animals. There is thus conceptual space between animal rights and veganism—there is at least one set of non-vegan foods that do not violate animal rights. These are products derived from animals that (not *who*) are non-sentient, and thus not rights bearers. There may be *moral* arguments against eating (products from) these animals—but this is not my primary focus. In this book, I am interested in just food systems, not morally proper diets. The state generally should not restrict people's choice to engage in (merely) morally improper behaviour; it should only restrict people's choice to engage in *unjust* behaviour.

Might there be reasons that producing food using non-rights-bearing animals is inconsistent with animal rights? Perhaps. Production methods may be unjust in the way that arable farming could be. For example, farmers might deliberately (and unnecessarily) kill rights-bearing animals for 'pest' control. Echoing the claims of Chapter 1, I seek foods that the zoopolis could (and perhaps should) include in the food system, rather than designing the all-things-considered *best* food system. That would be too great a task for a single book. I offer indications, though, of why I *suspect* that the all-things-considered best food system would include some animals: specifically, invertebrates.

Invertebrate-based foods are already familiar. 1,500 species of insects (this does not include *other* invertebrates) are traditional supplemental foods

(Boppré and Vane-Wright 2019, 47), while humans keep hundreds of species (trillions of individuals) captive today (Boppré and Vane Wright 2019, 31). We kill crustaceans, too, in eye-watering numbers. In 2008, 214 billion tiger prawns were among the 1,600 billion crustaceans we killed for food (Elwood 2019, 173). Invertebrates of *many* kinds appear in *many* cuisines. For example, people eat around 100 species daily in Japan (Pollo and Vitale 2019, 10)—but invertebrate foods are not 'only' Eastern.

At the time of writing, I live in York, a small British city. With a quick stroll, I am confident I could buy foods containing bivalves (mussels, cockles, oysters …); crustaceans (shrimps, crabs, lobsters …); cephalopods (octopuses, squids …); and bee products (honey, beeswax, royal jelly …). I would also find products containing invertebrates surreptitiously, such as sweets containing E120 (carmine), a colourant made of crushed scale insects. And that is just the start. In the right shop, snails, insects, spiders, and all sorts of sea creatures would not be out of the question.

Could invertebrates be part of a just food system? Perhaps we simply have an empirical question: Are these animals sentient? Or—as invertebrates make up 90 per cent of the animal diversity on earth (Horvath et al. 2013, 9)—*many* empirical questions. We cannot lump these animals together (Pollo and Vitale 2019, 7–8). But there are important philosophical questions, too. First, there might be reasons to doubt straightforward connections between rights and sentience. I put these questions aside. But there are also questions about what precisely is *meant* by sentience. I explore this below.

A vexing question comes when we are unsure of sentience. Sentient animals have rights, so warrant protections limiting how (if) we can use them for food. Non-sentient animals lack rights, so we can exploit them for food relatively freely. How should the state respond to the animals who do not obviously fall into either category? This is not, I argue, merely a pragmatic worry.

We can make a distinction between two senses of *rights*. First, there are rights possessed by animals, that justice demands we respect—the rights that, *if* these animals are sentient, they possess. But, second, there are the rights that the state should enforce. Normally, these are one and the same. But given that, in this case, we *don't know* what rights the animals possess, we cannot say that the state must enforce the rights they possess. We need an idea of the legal rights, if any, that the state should protect, given the uncertainty.

In this chapter, I propose a solution. We should treat some animals about whom we are uncertain as we should treat those beings we know to be sentient (i.e., extend rights to them). We should treat some animals about whom we are uncertain as we should treat those beings we know not to be sentient (i.e., not extend rights to them). And for some animals—about whose sentience we are *most* uncertain—we should protect only *one* right: the right not to have suffering inflicted upon them.

I offer example animals throughout. However, these examples come with a caveat: they are based upon my reading of the science. It is not my role to declare which animals are sentient. Such questions are partly philosophical, partly scientific. But line-drawing exercises are also political. *Any* proposal about inclusion and exclusion is going to involve line-drawing exercises. Current animal-welfare law does this—for example, EU animal-welfare legislation draws a line between cephalopods and other invertebrates. And such line-drawing is familiar territory for liberal states. Speed limits, ages of consent, tax bands—legislators must draw lines. (Such lines need not be *arbitrary*, though—they can informed, and can be democratically legitimate.) The categorizations I offer here are thus provisional and illustrative.

Non-sentient animals

For most animal-rights theorists, there is something 'special' about sentience (Cochrane 2018, 9).[1] Sentientists face critique from anthropocentrists, who call for a narrower scope of justice, but also from biocentrists and ecocentrists, who call for a broader one. Occasionally, too, they face critique from other animal rights theorists, who seek to include non-sentient animals in the scope of justice.[2] But here, we can follow the sentientist line.

There are several things we might mean when we call a being *sentient*, and it is important to distinguish them when addressing edge cases. What we can

---

[1] Tom Regan (2004b) appeals to the subject-of-a-life criterion as the foundation of rights, which is seemingly more demanding than sentience. I fear this was a misstep—both other theorists of old (e.g. Francione 2009, chapter 3) and new animal rights (e.g. Cochrane 2018) are sentientists. (Indeed, a focus on sentience, in the eyes of one commentator, is a close-to-defining feature of the new political camp. See Milligan 2015.)

[2] It is difficult to do this without opening the floodgates. Martha Nussbaum seems drawn to the idea that we owe non-sentient animals duties of justice as flourishing beings—although concludes that 'we have enough on our plate if we focus for the time being on sentient creatures' (2007, 362). One wonders whether non-sentient plants, too, could flourish (Nussbaum 2007, 447). It is difficult to see how notions of flourishing can include non-sentient animals in rights discourse without also including plants. If plants have rights, either we must conclude that we can kill rights-bearing beings for food with few worries, or we must embrace the conclusion that we should be eating (almost?) nothing. The former sounds very bad for animals; the latter, for humans. Eva Meijer, meanwhile, experiments with the inclusion of worms within the scope of justice (2019, chapter 6), regardless of their sentience, but can only do so by drawing upon a new materialist worldview, apparently undermining any initial commitment to animal rights. If Meijer commits to new materialism for the purpose of including duties to worms in her theory, why stop there? New materialists champion the agency not just of animals (sentient and otherwise), but of plants, soil, the weather, and more (compare Milburn 2019). But 'where every*thing* is due justice and respect, then nothing is' (Peter Gratton, quoted in Stanescu 2018, 230). Both Meijer and Nussbaum find themselves caught between a rock and hard place.

call *broad* sentience refers to subjective experience—phenomena, qualia, a point of view, *someone home*. There is *something that it is like* to be that entity (Nagel 1974). There is something it is like to be me, typing on a laptop—there is something it is like to be my dog, cautiously sniffing a bee. There is nothing that it is like to be the laptop. There *might be* something it is like to be the bee. Broad sentience differs from *very broad* sentience, which is the ability to have a non-phenomenological 'awareness' of the world—senses without experience. It is in this way that plants are 'sentient'; plants have apparatus for detecting light, water, gravity, and so on. But there is nothing it is like to be them.

In the *narrow* sense, sentience is the capacity to experience particular kinds of subjective experiences: states with *aversive* or *attractive* qualities. A narrowly sentient being does not *merely* have a point of view—she can have experiences that are positively or negatively valenced: states that she is (loosely) *happy* or *unhappy* about. Pleasure and pain are paradigm examples, but they are not the only such states—frustration and boredom are aversive but need not be painful; satisfaction and contentment are attractive but need not be pleasurable. There could be beings who experience aversion but not pain, or attraction but not pleasure. Thus, narrow sentience is not solely about pain. The sense of sentience concerned *solely* with pain we could call *very narrow* sentience.

These concepts are 'nested'. Very narrow sentience implies narrow sentience implies broad sentience implies very broad sentience. The inverse is not true. A being could, for instance, have subject experience without ever feeling, in the loosest sense, *happy* or *sad* about it, and thus be broadly sentient, but not narrowly sentient. This might sound like a mere thought experiment, but many invertebrates might be in this position.

Let us summarize:

> **Very narrow sentience**: The ability to experience pain.
> *… entails, but is not entailed by …*
> **Narrow sentience**: The ability to experience states that are positively and/or negatively valenced.
> *… entails, but is not entailed by …*
> **Broad sentience**: The ability to experience states.
> *… entails, but is not entailed by …*
> **Very broad sentience**: The ability to have some awareness of the world, even if that awareness is not phenomenally experienced.

It is not enough to say that sentience is necessary and sufficient for the possession of rights. We need to specify *which kind* of sentience.

If we frame rights as being based on the possession of *interests*, it is, I think, narrow sentience that is important. It is hard to see how beings who experience no aversive or attractive states could have things go better or worse *for them*, how they could have a *welfare*. Things may be better or worse for their 'functioning', or achieving the things they have evolved to do (the three *F*s: feeding, fighting, 'mating'). But, like plants, they do not care—if they did, they would be narrowly sentient, as *caring* involves aversion or attraction. The idea of interests presupposes a welfare (though not any particular *theory* of welfare). Subjective experience or otherwise, if an invertebrate lacks interests, they could not be a rights bearer for their own sake. There *is no* 'own sake'—there is nothing to protect.[3]

Thus, for animal rights, what matters is narrow sentience. Some invertebrates are narrowly sentient. There is a broad consensus, for example, that cephalopods can feel pain (sufficient, but not necessary, for narrow sentience). Legislators even enshrine this in law. Directive 2010/63/EU of the European Parliament—concerning the use of animals in science—restricts the use of vertebrates and cephalopods (but not other invertebrates). Like most scientific consensuses, there is no *universal* agreement; for example, at least a few scholars doubt *any* animals are sentient. But for our purposes, this is sufficient. Octopuses, squid, cuttlefish, and the rest are off the menu. (In later chapters, I explore plant-based and cultivated meats—so there may be a place in the zoopolis yet for lovers and producers of the Galician *polbo á feira*, the Turkish *ahtapot salatası*, and the Japanese *takoyaki*.)

There are also invertebrates who are *not* narrowly sentient. A go-to example in the invertebrate welfare literature is jellyfish. There is basically no reason to believe that jellyfish are sentient; one would struggle to find a scientist seriously defending the possibility. So, jellyfish firmly *are* on the menu: there is no in-principle reason that the zoopolis should ban their consumption (or harvesting, farming, processing, sale ...), just as there is no in-principle reason that it should ban the consumption (or ...) of carrots.

This conclusion might sound hollow. It is not. First, although perhaps unfamiliar to some readers, there are jellyfish dishes—even delicacies. *Haeparinaengchae* is a Korean salad with jellyfish; *yangjangpi* is a Korean-Chinese appetizer with jellyfish. The animals are eaten as tempura or sushi in Japan,

---

[3] Andy Lamey (2016) argues that non-valenced subjective experience is enough—but, I think, he goes wrong. While I cannot offer a full response here, I note that his argument depends upon contrasting the treatment of a *necessarily* neutral being (who is broadly but not narrowly sentient) with a being who is *contingently* neutral (who is narrowly sentient but never experiences valenced states). This comparison is not compelling; a narrowly sentient being who has never experienced valenced states still has interests concerning them, just as a new-born whale yet to surface has interests in access to air. A broadly sentient being who is not narrowly sentient has no such interests.

or as street food in Thailand.[4] Second, jellyfish are an example from the invertebrate welfare literature, rather than (solely) a concrete proposal for an element in the food system. There may be *other* invertebrates to which we can reliably attribute non-sentience. Some of these may be more familiar to western diners. (More on this later.)

Although significant, the fact that non-sentient animals could be part of the food system is an easy conclusion, theoretically speaking. The trickier case concerns animals who *may* be sentient.

Uncertainty about sentience: The problem

On invertebrate sentience, there is considerable uncertainty (Elwood 2011; Horvath et al. 2013; Adamo 2016). Partly, this is due to the difficulty of the questions involved. However, it is also due to the lack of interest among animal welfare scientists. There is little information available on invertebrate sentience/welfare (Carere and Mather 2019, 1), or, indeed, invertebrate farming methods (Boppré and Vane-Wright 2019). Applied work in 'bug ethics' must deal in uncertainties.

These uncertainties might seem decidedly pragmatic—invertebrate welfare science will chip them away in time (Milburn 2015, 456). Thus, perhaps the present work is the wrong place for these problems. From a principled perspective, perhaps the matter is simple: Sentient animals are off-limits, non-sentient animals are not. But this is over-simple: these may be matters that we *shouldn't* resolve, or that we *can't* resolve.

Perhaps we should not endorse the work necessary to pursue invertebrate welfare science (Marks 2017, 2). Searching for aversive responses involves vivisection of the most troubling kind. Reviewing work on invertebrate welfare, I have read reports of crushed or severed limbs; acid rubbed on sensory organs; electrocution; 'learned helplessness'; and more. Although the claim is beyond the scope of this work, these are perhaps not things the zoopolis should endorse, encourage, or (even) permit. Without changed methods, searching for sentience necessitates literal torture.

Even with a torture-free route to rigorous sentience research, a second problem looms. Perhaps the questions are irresolvable, with uncertainty all the way down. Another's subjective experiences are inaccessible to us. This

---

[4] To add to my comments in the introduction of this chapter, I was intrigued to find jellyfish in a Chinese supermarket a mile from my house in York. This is despite regularly encountering disbelief among Brits that jellyfish are edible.

is not philosophical handwringing, but a core problem acknowledged by biologists (Carder 2017, 3).

There is also no agreement on what evidence suffices to attribute sentience (Elwood 2019), *especially* invertebrate sentience. The deck is stacked against invertebrates, as welfare scientists base ideas about sentience attribution on vertebrates, whose sentience may differ from invertebrates'—evolutionarily, phenomenologically, physiologically. Indeed, we may need to rethink our ideas of subjective experience altogether. Perhaps, for instance, insects are broadly sentient but not narrowly sentient (see Lamey 2016)—a phenomenology difficult for us to imaginatively inhabit. While the relevant experts may reach a consensus, it is more likely, I think, that disagreement will persist.

Currently—and foreseeably—we cannot be *certain* that any other animals (including other humans) are sentient. Solipsism (the belief that only one's own mind exists) remains a defensible, if unpopular,[5] position. But we may never even have *relative* certainty about some invertebrates. But their normative status warrants consideration.

That said—although acknowledging concerns about vivisection's injustice—the difficulty of ascertaining sentience does not mean we should not try. We have good reason to explore the minds of any invertebrates exploited for food. It is important to *check* if we are violating rights. For example, there is a wrong if, in war, we do not check whether a house we are about to bomb contains enemy soldiers (who are legitimate targets) or civilians (who are not). Crucially, Cecile Fabre (2022, 62) argues, we wrong *those in the house* by failing to check. This is true even if they are, as it happens, enemy soldiers. Failing to check introduces an injustice when none exists (i.e. soldiers present) and compounds an injustice when one does exist (i.e. civilians present). Similarly, failing to take reasonable steps to ascertain if an exploited animal is sentient could compound injustice if the animal is sentient, and maybe *introduce* injustice *even if* they are not sentient.[6]

Nonetheless, uncertainty will likely remain. It is thus worth distinguishing three broad categories of beings. First, there are what we might call *Probably*

---

[5] There is only one solipsist, as the joke goes.

[6] But, it could fairly be contended, if the animals are not sentient, we cannot wrong them. In an earlier draft of her book, Fabre (private correspondence) explored a case in which the staff of a target building may be civilians or may be robots. She writes that, although the robots lack moral status and so soldiers cannot owe them a duty to check, 'it is hard to let go of the thought that [the military] *ought* to have checked, just in case. It seems that the only way to make sense of that thought is by accepting undirected duties—duties, that is, which one does not owe to anyone'.

Sentient animals. These are beings about whose sentience (outside philosophy seminars) we are basically certain, *and* those beings about whose sentience we have *some* doubt, but sufficiently little that we should happily label them sentient in the eyes of the law.

Second, there are those beings at the other extreme. As panpsychism—the belief that all physical entities have some form of consciousness—has at least a minimal plausibility,[7] we can conclude that we are not *entirely* sure about the *non*-sentience of *any* material thing. Nonetheless, we can characterize as *Probably Non-Sentient* those beings about whose non-sentience (outside philosophy seminars) we are basically certain. (Technically, any animal or non-animal labelled non-sentient belongs in this category.)

There is a deliberate asymmetry between the 'basic certainty' required for Probably Non-Sentient and the 'sufficiently little doubt' required for Probably Sentient. Although we should not dismiss the negatives associated with positive sentience attributions (more on this later), there is an asymmetry of risk (Sebo 2018, 58–60). While humans certainly lose much in the closure of (say) shrimp fisheries, the basic rights of animals, even if we are not *completely* certain that these animals have rights, carry greater normative weight.

This leaves a third category: the *Plausibly Sentient*. These are animals about whose sentience we are sufficiently uncertain that we should not be confident calling them Probably Sentient or Probably Non-Sentient—the uncertainty in their case warrants treatment unlike the treatment offered to (Probably) sentient *and* (Probably) non-sentient animals.

These categories are, I admit, hazy. But, again, the line-drawing here is not a solely philosophical task, and may require political decision-making. I am not, for example, going to say that we need to be x% sure to label a being Probably Sentient.

Before I explore these categories, I acknowledge a challenge. Why three categories? Why not say something like this: If we are $\geq$x% sure that a particular animal is sentient, we should treat that animal as if she is sentient; if we are <x% sure that an animal is sentient, we need not (Fischer 2016a, 259)? The trouble is that this proposal commits us to treating animals with an x% chance of sentience *very* differently to animals with an (x−1)% chance of sentience. While legislatures should indeed draw lines, it is worth exploring whether more subtle gradations might better allow laws to approximate moral truth.

---

[7] I am prepared to grant this given that cleverer people than I have defended it.

Uncertainty about sentience: Possible solutions

Establishing if Plausibly Sentient animals belong in our food system is an important task for animal ethicists. Several strategies present themselves. It is my claim that the most compelling approach will combine these strategies.

Strategy one: Don't know, don't kill

The first option is *don't know, don't kill* (Guerrero 2007; Singer 2015, 103; Sebo 2018, 58–59). This is a precautionary principle: if we are unsure about a being, we should treat them as if they have moral status. That is, when we have beings who could be sentient, we should treat them as if they are sentient (Fischer 2016a, 259; Jones 2017, 2), and thus rights bearers. Precautionary principles are popular (e.g. Fischer 2016a; Birch 2017a; Pollo and Vitale 2019, 17–18), and something like *don't know, don't kill* is implicit among vegans abstaining from invertebrate-based food. However, we should reject a straightforward *don't know, don't kill*.

First, it forces vegans to bite unappetizing bullets. It is minimally plausible that plants are sentient; indeed—recall the panpsychist—it is minimally plausible that *bullets* are sentient. And this creates a problem for *don't know, don't kill* (Sebo 2018, 57). If we treat plants as if they are sentient (thus, rights bearers) we will have *nothing* to eat—or must commit to eating rights-bearers. *All* entities being rights bearers sounds like *no* entities being rights bearers (Stanescu 2019, 230–232).

Second, *don't know, don't kill* underestimates the harms in misattributing sentience in politics. In extending rights to edible invertebrates, we adversely impact many people (Adamo 2016, 78; Birch 2017b, 11). As I explored in Chapter 1, this might include (though is not limited to) people who work with these animals/foods, foodies, and people from whom this food is culturally/personally important. And (as, again, I noted in Chapter 1) it may necessitate a switch towards forms of plant-food production that are more environmentally harmful and/or more harmful to animals than the production of invertebrate-based food (cf., e.g., Meyers 2013; Fischer 2016a). Troublingly, precautionary principles pull in both directions. If we declare these animals sentient, we risk harm to humans; if we declare them non-sentient, we risk harms to *them* (Klein 2017, 1). *Don't know, don't kill* places all the risk of harm on the heads of the humans. As a political principle, then, it is unjustifiably misanthropic.

Third, *don't know, don't kill* runs into difficulties when applied beyond narrow food-related questions. Perhaps 'minilivestock' should be the *least* of our worries if we are concerned about harm to invertebrates (Boppré and Vane Wright 2019, 24). Some rough calculations suggest cars (alone) on Dutch roads (alone) kill 800 billion insects (alone) *every six months* (Fischer 2016b, 2). And let us think about arable agriculture. The mind boggles when one thinks of the invertebrate animals killed by harvesting and pesticide use. A hectare (a 100m-by-100m square) of agricultural land contains 250 million to over a billion insects. (Plus *other* invertebrates.) If we consider these insects alongside vertebrates killed in arable agriculture, growing wheat starts to look *deeply* questionable (Fischer and Lamey 2018, 419–420).

In 2017, the United Kingdom had over 17 million hectares of agricultural land; the United States had over 405 million.[8] A conservative estimate gives over 4 quadrillion insects on agricultural land in the United Kingdom; a higher estimate gives us 17 quadrillion. In the United States, the lower estimate gives us over 101 quadrillion insects, and the higher estimate gives us 405 quadrillion. (405 quadrillion insects is over 50 million insects per living human. It is over 4 million insects per human who has ever lived.) It is hard to know how to start thinking about these numbers, but they do suggest that *don't know, don't kill* forces us to make *massive* changes to our lives and societies in insects' name.[9] Maybe we should. But for lots of us, I think, this *rebugnant conclusion* will show that we have gone wrong.[10]

Don't know, don't kill does not work as a general principle for animals about whose sentience we are uncertain, nor (specifically) Plausibly Sentient animals. But it *is* appropriate for Probably Sentient animals. I include all vertebrates (including fish) in this category, subject to a few assumptions (living animals, post-natal animals, non-brain-dead animals, etc.). Among invertebrates, subject to the same sorts of assumptions, I include cephalopods and decapod crustaceans.[11] We should treat these animals, even if there remains some doubt, as if they are sentient, and thus afford them rights. Consequently, it is (on the face of it) appropriate that the state prevents people from farming/harvesting these animals.[12]

This represents a not-insignificant loss for humans. Take shrimps. In protecting them, we lose a food source—the Norway lobster, beloved by

---

[8] See http://www.fao.org/faostat/en/#data/RL.
[9] As anyone with a compost heap knows, this problem is not unique to industrial arable agriculture.
[10] I borrow the pun from Jeff Sebo (2022, 175–178); he uses it in a not-unrelated context.
[11] My inclusion of decapod crustaceans on this list is controversial, but consider prawns (shrimps). Humans target several species of prawn in fishing and aquaculture, but (absent clear arguments to the contrary) it makes sense to consider decapod crustaceans together. A range of tests have indicated, though not proven (Diggles 2019), that decapod crustaceans experience aversive phenomena—e.g. pain (Elwood 2019). Glass prawns, for example, groom antennae or eyes for an extended period when biologists apply acid, but not when they apply seawater (Elwood 2019, 159).
[12] There are, as argued in the rest of the book, just ways to acquire the meat of and/or farm rights-bearers.

foodies, is off the table; families who start their Christmas dinner with a prawn cocktail must adjust traditions. Our cultural identities are impacted—no more Morecambe Bay shrimp for Lancastrians, while seafood Gumbo (Louisiana's state dish) needs new ingredients. And people who find valuable work harvesting, preparing, shipping (etc.) shrimps have lost something. A ban disproportionately impacts marginalized, deprived people who fish for subsistence—including communities in south-east Asia and Africa—raising food justice concerns.

However, there are also important reasons that the animal-rights perspective aligns with a broader food justice perspective. Commercial shrimp fishing is environmentally destructive, with damage to the seabed and huge 'bycatch'. The industry is also associated with contemporary slavery. Shrimp farming raises its own problems. Aquaculturalists feed animals to carnivorous crustaceans—raising concerns about the other animals' rights, environmental impact, and so forth. This is not to say that all shrimp/prawn harvesting raises these concerns. But it *is* to say that the end of shrimp fishing/farming would not be an unmitigated disaster for humans.

Strategy two: Don't know, don't worry

Mirroring *don't know, don't kill* is what I call *don't know, don't worry*: if we do not know whether a being is sentient, we should treat it as if it is not (Sebo 2018, 55–56). This 'incautionary' principle may sound implausible. However, it has historically been the consensus among policymakers, scientists, and philosophers (Jones 2017, 2), and it seems to remain the position of some welfare scientists (Wadiwel 2016, 2). I think there is something to be said for it. We should not be too comfortable deploying the state's coercive power against people when they may be doing nothing wrong.

That said, I reject *don't know, don't worry* for reasons mirroring challenges to *don't know, don't kill*. First, given the uncertainty of sentience attributions, the principle means we cannot affirm *any* attribution of sentience—we could not recognize *any* animal as a rights bearer. Some may be sympathetic to this outcome. But, again, I am not *certain* of *your* sentience. That way, anarchy—in its most pejorative sense—lies.

Second, *don't know, don't worry* shows callousness towards the (potentially rights-bearing) animals it leaves behind. To return to some earlier-cited statistics: we kill over 200 billion tiger prawns annually. These are, I think, Probably Sentient—but their non-sentience is possible. Are we ready to adopt a principle whole-heartedly supporting an industry potentially killing 200 billion rights-bearing beings annually?

Don't know, don't worry, like *don't know, don't kill*, is useful when dealing with *some* entities about whose sentience we are uncertain—specifically, in this case, Probably Non-Sentient entities. It is appropriate that we permit individuals to use *these* beings freely, even if we are not *absolutely* certain that they are non-sentient. In this category, in addition to non-living things (vitamins, minerals …), I include plants (vegetables, fruit, crops …), fungi (mushrooms, yeast, moulds …), and other non-animals (e.g. kelp). Crucially, I also include some animals. Jellyfish are the example I have already given. The state has no business preventing these things being part of the food system (Or, if it does, this is not because of these things' rights.)

To demonstrate the significance of this conclusion, consider the bivalves eaten as 'oysters'. There is little evidence that these animals are sentient. They possess no brains, a very simple nervous system, and limited senses. They are mostly sedentary, meaning sentience—a biologically demanding capacity— would have little evolutionary value for them. If a demanding capacity serves little purpose, evolutionary pressure acts against its development. Although little-discussed in the animal-rights literature, the merits of eating oysters are championed by some animal ethicists exploring the ethics of eating (Jacquet, Sebo, and Elder 2017; Budolfson 2018), and 'ostroveganism'/'seaganism' has gained some real-world traction.

Despite the lack of evidence, at least *some* defend the claim that oysters are sentient (e.g. Marks 2017). But there is a sufficiently low chance of sentience that we should not be affording legal rights to these animals. Recall that this means deploying coercive power—and recall that we should be cautious about the state overstepping its mark, especially when there are other important values at stake. And there are. Jennifer Jacquet, Jeff Sebo, and Max Elder, although suggesting that plant-based food would be preferable to bivalve-based foods, write that:

> Of all the aquatic animal species groups that we eat as food, bivalves appear to be the most promising in terms of minimizing ecological harm (in some cases they may even be beneficial), minimizing food security harm (as highly nutritious organisms that do not rely on outside food sources), and minimizing animal welfare concerns related to captive rearing.
>
> (Jacquet, Sebo, and Elder 2017, 32)

I would go further, and say that there are good reasons to believe bivalve-based foods might often be all-things-considered *preferable* to plant-based foods (Milburn and Bobier 2022, 6). As noted, filter-feeding bivalves may be environmentally beneficial. Meanwhile, some forms of 'plant'-based farming might be very *bad* for the environment (cf. Meyers 2013). Compared to gathering kelp, for example, oyster farming seems to have compelling environmental credentials (Kravitz 2018).

And might there be benefits to other animals from oyster farming? Perhaps—if harvesting oysters involves less harm to sentient animals than harvesting plants (cf. Fischer 2016a, 257–258), which there is every reason to think it might, given the harms to animals in plant agriculture. (This includes vertebrates, and, as noted, insects—which are invertebrates more likely to be sentient than oysters.) There are also benefits to human consumers. Not only might access to another food source be important for food security—as Jacquet, Sebo, and Elder note—and for responding to world hunger (Meyers 2013, 124), but oyster farming opens many routes to the good life. Oysters are part of coastal cuisines; the industries directly or indirectly exploiting oysters provide good work; and oysters are a foodie favourite.

Thus, not only does the zoopolis lack compelling reasons to restrict oyster consumption, but it has reasons to champion the oyster industry—even over a plant-based food system.[13]

Strategy three: Gradations of precaution

In the shadow of the inadequacies of *don't know, don't kill* and *don't know, don't worry*, we find a range of principles in the middle—affording *some* protection to Plausibly Sentient animals. The challenge is mapping this middle way.

We can reject some routes. A common strategy is an *expected utility* (or *expected value*) account (Sebo 2018). An expected utility approach would involve multiplying the bad that *would* be done to the animals by the farming methods if the animals were sentient by the likelihood that the animals are sentient, and comparing this to the bad that would be done to humans if the practice was banned. The 'less bad' outcome is the one favoured. For example:

> An industry farms 5 million invertebrates with a 1% chance of sentience. The farming method would impose five 'disutils' on each if sentient. 5 million by five by 0.01 gives 250,000 disutils. In comparison, 5,000 people relying on the industry would

---

[13] There may still be *moral* reasons to avoid oyster consumption. Jewish dietary laws of *kashrut* prohibit the consumption of shellfish, and Rastafari and Jains typically favour vegetarianism. Even if the zoopolis permitted oyster consumption, those believing it immoral would be welcome to abstain and proselytize.

experience forty-five disutils if it was banned. This totals 225,000 disutils. Banning thus results in higher expected utility than permitting.

There are two sets of problems with this approach—one concerning calculation, one normative. The calculation problems are clear: We have no way to assign these numbers. What does 'five disutils' look like? And how could we agree that there is a '1% chance' that an animal is sentient? We must draw lines on any position—but calculations like this demand precision we cannot have. Even if we could get these numbers, the calculations quickly become *very* complicated. What effect will the ending of this farming have on local and not-so-local ecosystems? Economies? Cultures? How do we account for the animals who will no longer be born?

Perhaps we can overcome these problems. But the normative challenge speaks to the unsuitability of these calculations for an account of *justice*. Simply put, we should not be reducing lives—human or animal—to mere 'containers' of value and disvalue. Nor should we be weighing lives like bonbons on a cosmic scale. Treating animals (including us) like this is *disrespectful* (Regan 2004b). There *may* be room for expected value reasoning for *moral* purposes.[14] But this disrespect is more worrying when it comes to questions of policy.[15]

Can we capture the appeal of an expected value approach without these challenges? One possibility would look something like the following.

> In cases where a particular individual is Plausibly Sentient, and treating that individual as sentient wouldn't prevent us from fulfilling our obligations to any being that is Probably Sentient, we ought to treat that individual as though it is sentient.
>
> (adapted from Fischer 2016a, 260)

---

[14] Jeff Sebo offers the following compelling example:
> suppose that you think there is, say, a 12 percent chance that a lobster (hereafter bio-lobster) is sentient, and you think there is, say, an 8 percent chance that a functionally identical robot lobster (hereafter robo-lobster) is sentient. Now suppose that a house containing both lobsters is burning down and you can save either but not both. Assuming you should save one of the lobsters, which one should you save? ... while both of these lobsters *might* be sentient, the bio-lobster is also a bit more *likely* to be sentient than the robo-lobster is, given your evidence, and therefore you morally ought to break the tie in favor of saving the bio-lobster, all else equal. (Sebo 2018, 58)

Sebo surely gets something right here—but this is not a policy question. One does not need to embrace expected value approaches to recognize the usefulness of this kind of 'tiebreak' reasoning.

[15] Precautionary principles, popular in policy discussions, are sometimes conflated with expected value approaches, but advocates of the former in conversations about invertebrate sentience are quick to separate the two (Fischer 2016a, 260–261; Birch 2017b, 11).

On a rights framework, this would (troublingly) seem to become all-or-nothing. As outlined in the previous chapter, there are lots of reasons for states to endorse animal agriculture, even if other considerations may often override them. This includes farming Plausibly Sentient animals. Take apiculture. If reasons in favour of state support (or tolerance) of a honey industry are sufficiently strong that treating bees as sentient—thus banning apiculture—would mean the state was failing to fulfil obligations towards beekeepers, then all apiculture would be acceptable. If these reasons are not sufficiently strong, however, then all apiculture would be unacceptable. Indeed, perhaps the reasons in favour of state support/tolerance of apiculture are sufficiently strong to suggest that treating bees as sentient would wrong apiarists *if* bees are not sentient, but *not* if they are sentient. But we do not know. That is the point. In this case, the principle is unhelpful.

It is, however, a step in the right direction. What we need is something that can offer gradations of precaution, but that does not result in an all-or-nothing approach.

Strategy four: Rights-lite

We could begin by asking *which* interests particular invertebrates (may) possess (Fischer 2016b, 4). Affording invertebrates rights may be less burdensome if we do not much impact their interests.

Relatedly, we could adopt a graduated account of sentience. This amounts to the same thing. Sentience is binary, but there can still be incredible *diversity* in sentience. Jonathan Birch, who proposes a 'sliding scale' of sentience, concedes that his sliding scale is compatible with fact that sentience is binary: 'even if the presence or absence of feeling is a binary property, the forms of animal feeling can vary by degrees along multiple welfare-relevant dimensions' (2018, 2). We do not need to pretend that sentience comes in degrees; we just need to recognize that there are normatively pertinent differences in the aversive and attractive states accessible to sentient beings, giving rise to different (strengths of) interests.

What interests *would* Plausibly Sentient invertebrates have *if* they were sentient? This is tricky. We are not certain if these animals are sentient, and we are not certain what their phenomenology would be if they *were* sentient. And invertebrate diversity means different invertebrates may have different (in Birch's language) 'levels' of sentience. Nonetheless, we can make some general observations.

In short—and to look ahead slightly—my contention is this: It is sufficiently unlikely that these animals would have some potential interests, and/or sufficiently likely that *if* the animals had those interests, they would have them only weakly, that it might be appropriate for us to distinguish between interests when assigning legal rights. Or, to put it another way, perhaps we should adopt a precautionary principle when it comes to *some* potential interests, and thus protect them with rights, but adopt an incautionary principle for others, and thus not protect them. But which interests am I talking about?

Sentient invertebrates have an interest in avoiding pain. Or, more precisely, they have an interest in avoiding aversive experiences. (Pain may not be among them—or may not be the most important of them.) Similarly, they have an interest in experiencing pleasure (or other attractive states). The interest in avoiding aversive states may be very strong. Indeed, it is hard to imagine what interest could be stronger. For reasons explored shortly, the interest in experiencing attractive states may not be *as* strong.

What about death? The relative unsophistication of invertebrate psychologies is significant. First, invertebrates will likely have no conception of themselves as a continuing entities; they will not recognize themselves as the same being from time to time. Second, they will have no fear of death—even if they might 'fear' aversive states, they cannot comprehend the end of their own existence, and so could not fear it. Third, they will have no plans, projects, or goals scuppered by their death. On some accounts of the badness of death (e.g. Belshaw 2015), these features mean the animals have no interest in continued life.

On *deprivation* accounts of death's badness, the bad of death is that it deprives us of good things we would have had. Such accounts could apply to invertebrates (Hadley 2019b). However, we must ask what the invertebrate would lose. Drawing upon biocentric notions of 'biopreferences', John Hadley (2019b) suggests that insects have 'preferences' *qua* living beings that are frustrated by death. Biocentric accounts, however, open the floodgates. Not only do they extend interests to sentient invertebrates, but they extend them to non-sentient animals and other living things. If plants *and* animals have rights, we have a problem—as noted earlier.[16]

---

[16] Hadley points towards, but does not develop or endorse, a mixed animal-rights/biocentric approach in which the possession of rights depends upon the possession of certain psychological features (e.g. sentience) but the content of those rights could draw upon biocentric ideas about interests. This seems arbitrary. If biocentric interests are genuine interests, we should respect them, wherever they are found.

Let us try another route. Death scuppers any chance of future attractive experiences for invertebrates. However, even assuming sentience, an interest in accessing these experiences would be weak—much weaker than an interest in avoiding aversive states. As above, Plausibly Sentient invertebrates have limited connection to their future selves, if any. What this means is that, despite the *physical* continuity of their body, they have little *psychological* continuity over time (McMahan 2002). There is only a loose sense according to which we can call the simple being *the same* being at one time and another. Consequently, present pain (or an otherwise aversive state) is a lot worse than future pain (or …) *for this being*, and present pleasure (or an otherwise attractive state) is a lot better than future pleasure (or …) *for this being*.

It also means that it is hard to outweigh present pain with future pleasure. It is reasonable for someone to choose for me to undergo a painful procedure now to spare me pain in the future. (Assume I am somehow incapacitated, hence someone else making the decision on my behalf.) The future me is still *me*. It is less reasonable to make this same choice on behalf of a psychologically simple sentient being. A future being will not experience pain that they would have done, but it is less clear that it is the *same being* as the one who *does* undergo pain in the present.

Troublingly, this lack of psychological complexity may make pain (or other aversive states) *worse* for the animal in question than for us. They cannot reason pain away; they cannot recall or look forward to a time of non-pain; their *whole universe* becomes pain (Rollin 2011, 431; Garner 2013, 126). These factors serve to justify the intuition that it is worse to torture a relatively psychologically unsophisticated animal than to painlessly kill her, while the opposite may be true of more complex beings.

Claims about psychological continuity are contentious. While I think the lack of continuity salient, I allow that some will not. Even if we ignore psychological continuity, however, we might have reason to worry that the losses experienced by Plausibly Sentient invertebrates (if sentient) upon death are not substantial. First, invertebrates typically live short lives, with few positive experiences. They are thus not losing *much*. Second, their lack of psychological sophistication may mean that the intensity of their pleasures (or other attractive states) is much lower than the intensity of the pleasures (or …) of more sophisticated animals. The pleasures accessible to these animals are simpler and less varied. (Incidentally, there may be an asymmetry here between positive and negative experiences. Pain is a simple experience—but a bad one. Perhaps simple animals experience terrible negative states, but only mild positive states.) In all, we have reason to expect that invertebrates' interests in continued life are weaker than those of many animals, and weaker than their own interest in avoiding aversive states.

Might another account of death's badness conclude that invertebrates—if sentient—have an interest in continued life *equal* to other animals' interests in continued life? Theorists might aim to resist claims about interests' strength by arguing that all sentient animals are equal. But the idea that all (sentient) lives are equal is consistent with the claim that beings can have a greater or lesser interest in continued life. The same applies in the human case. When providers make healthcare decisions based on Quality-Adjusted Life Years, they do not claim that the lives of impaired, elderly people are worth less than the lives of young, healthy people—they simply observe that the death of the elderly person means less of a loss for the elderly person than the death of the young person means for the young person.

I conclude that *if* invertebrates are sentient, they likely have a relatively strong interest in not suffering, but a comparatively weak interest in continued life. My proposal, thus, is this: We extend to Plausibly Sentient animals *one* legal right: *a negative right against the infliction of suffering*.

Let us unpack this. This is a *legal* right in the sense that I am not committed to the claim that this is the only right that, *as a matter of justice*, we owe them. Nor am I committed to the claim that we *do* owe them this right *as a matter of justice*. However, it is, I propose, the best way for legal systems to put into practice the *possible* rights owed to these animals as a matter of justice. (I add, incidentally, that we need to think about how to inscribe this right into law. I suspect we would best realize it by encoding a range of specific rules. These would not be unlike existing animal welfare legislation, although they would be firmer.)

Suffering is here understood broadly to encompass all aversive states accessible to particular sentient beings—not just pain. Suffering is thus any enduring, intense aversive state. The right protects beings from treatment that (our best estimates suggest) would inflict suffering (understood broadly) if these beings were sentient. A starting point is that electrocution, burning, and so on are illegitimate. But there may be things deeply aversive to these animals that are unobjectionable to us—for example, light (see Carder 2017)—and vice versa. Many invertebrates will thrive in conditions that would represent the worst excesses of industrial agriculture if experienced by vertebrates: hot, stinking, with animals crawling over each other.

Why a *negative right*? A *right* is stronger than a call for improved welfare; a right *precludes* certain treatment.[17] But this is a *negative* right because it is a tool limiting (potential) wrongs on these animals in pursuit of food, not a tool to force us to embark on a quest to alleviate the suffering wild invertebrates. If we posit a *positive* right to assistance in cases of suffering, however, this may be what we are required to do. All suffering of wild invertebrates would be an injustice that we should remedy, and, given the sheer numbers involved (recall my back-of-a-coaster calculations about insects on farmland), we could easily argue that it would be among the most urgent of injustices.

But framing this as a negative right is not an ad hoc stipulation. Rights theorists recognize a moral distinction between acts and omissions—that I would violate someone's rights by killing them does not entail that I violate their rights by failing to prevent them from being killed.[18] I violate the rights of a pheasant if I shoot her for my table, but do not necessarily violate her rights if I fail to assist her when she is taken by a fox for (whatever passes for) a fox's table. So too with Plausibly Sentient animals. We violate their (legal) rights when we (do things that would) inflict suffering upon them, even while we do not when we fail to assist them when they suffer for non-anthropogenic reasons.

And why suffering, rather than any other (potential) interests? As argued, an interest in not suffering (understood broadly) is the most significant possessed by the Plausibly Sentient. At least, it is more significant than an interest in staying alive, and it is tricky to imagine what other interest would be stronger. Thus, protecting these animals from suffering protects them from the most egregious injustices they would face if sentient (cf. Garner 2013). Meanwhile, the focus affords recognition to other values associated with using these animals. Such values are not sufficient to override the demands of justice. But they are sufficiently strong for us not to be positing legal rights without good reason. And the *small* chance that these animals have a *small* interest in continued life does *not* seem a sufficiently good reason to posit a right that will remove good things for humans—and, perhaps, good things for

---

[17] Even if rights are the tool for dealing with *real* interests, are they for *possible* interests? I think so. Consider the case of property of a missing person. The property remains theirs while missing—a stranger would violate the missing person's rights by helping themselves—until the point when the line is crossed, and the person is declared gone. We do not know if the missing person has any interest in their property. But we presume rights while sufficient uncertainty remains.

[18] Old abolitionist animal-rights theory focused on negative rights (Milligan 2015). New political animal-rights theorists address the positive rights too, but they are judicious—while negative rights are universal, positive rights are (generally) not.

other animals and the environment. While Plausibly Sentient animals' legal right *restricts* humans, it does not rule out eating or farming Plausibly Sentient animals, provided such farming does not involve practices that would inflict suffering.

Interestingly, the farming of Plausibly Sentient animals may already be high welfare. Take insect farming. First, farmed insects favour high population densities (Meyers 2013, 124); second, the lack of medical care means farmers must prevent disease, or risk losing investments (Boppré and Vane-Wright 2019, 39); third, environmental controls must be precise (Boppré and Vane-Wright 2019, 37, 39).

These factors mean that environments optimal for insect welfare and environments optimal for farmers' profits coincide. That said, I do not want to romanticize invertebrate agriculture, and fully accept that some forms of invertebrate farming will involve a great deal of suffering—assuming sentience (Carere and Mather 2019, 4). Getting to the bottom of which practices are consistent with the approach defended here, and which are not, requires empirical data to which, at present, we have limited access.

As an illustration, take apiculture. Bees are insects. As with decapod crustaceans (but not bivalves), there has been some modest research on insect sensation. But, even when compared to decapod crustaceans and bivalves, there is little consensus concerning their sentience. On the one hand, Colin Klein and Andrew Barron (2016) defend insect sentience, prompting dozens of (often sympathetic) scholarly responses.[19] On the other hand, Claudia Garrido and Antonio Nanetti (2019) publish a chapter on bees in a book about invertebrate welfare, but do not even acknowledge the possibility that bees are sentient. Bees seem to be a prime candidate for the Plausibly Sentient category. Based upon the current science, there is a non-negligible, but not overwhelmingly significant, chance that these animals are sentient. Thus, I propose that we appropriately protect them with one legal right: a right not to have suffering inflicted upon them.

We might think that the zoopolis would still ban apiculture. Why would the zoopolis risk beekeepers inflicting suffering if bees have a right against said infliction? I reject this possibility. If apiarists can gather honey (wax, royal jelly …) without doing things that would inflict suffering upon bees (if sentient), the zoopolis should let them do this. Not only does the state risk overstepping its mark in banning apiculture, but the loss of beekeeping would be regrettable.

---

[19] Although precisely what kind of sentence they are arguing for is an open question (Lamey 2016).

Why? Let us explore some specifics. First, apiculture is something central to understandings of the good life. Richard Taylor was an American philosopher who happened to write both academic texts on the meaning of life and popular texts on beekeeping. In his *The Joys of Beekeeping*, he wrote:

> One's happiness is, of course, something personal, something more his own that any possession. We do not all find it in the same way. Some never find it at all. Possibly most never do, even when the means are at hand. But I have found my bees and all the countless things I associate with them a constant and unfailing source of it. … Without bees my own existence would be a shadowy thing like a world without flowers or without stars or without the song of birds.
>
> (Taylor 1974, 4–5)

For Taylor, the good life is the life of the apiarist. Access to honey, similarly, gives some people's lives meaning. When he heard that I did not eat honey, a former officemate of mine said, with all sincerity, that honey on Greek yoghurt was (for him) one of life's great pleasures. Or consider followers of Heathenism, a new religious movement. For Heathens, mead (a drink made with honey) is of central significance. Heathens drink mead during the *symbel* and offer it during the *blót*—two Heathen rituals. Taylor, my former colleague, and Heathens all lose something of central value if animal rights rule out apiculture. Now, *if* animal rights mean that the state *must* rule out apiculture, so be it. But we should oppose banning apiculture if a ban is not necessary.

Could apiculture be legal? Close, rigorous examination of assorted approaches to beekeeping would be worthwhile—but beyond the scope of this chapter. But I can offer some indications. Some apicultural practices would be permissible. Beekeepers could take honey, *providing* replacements did not lead to suffering. They could kill 'underperforming' or 'excess' queens (or whole hives), *provided* it was painless. They could purchase bees, *provided* the housing and transport practices did not entail suffering. Some other practices would be impermissible. Beekeepers could not clip queens' wings—if bees are sentient, wing removal will probably involve suffering.

There are, however, questions that are trickier. Apiarists use smoke to subdue bees. Does this induce an aversive state in the bees? Even if not painful,[20] it is *possible* that non-pain aversive states are very bad for these animals. And might bees have hive-level sentience? I am sceptical. But it is not for me to prejudge the results of further study.

---

[20] Smoke blocks bees' sensory apparatus. It also changes bee behaviour, which may indicate aversion—but probably not pain. (Aversion to fire seems to be a useful trait to evolve.)

Alternatively, we could (in time) conclude that bee sentience is sufficiently likely that bees are Probably Sentient, and thus owed more rights—or sufficiently unlikely that we owe them no rights (Milburn 2015, 456). Full rights would presumably mean the end of apiculture—an unhappy result for Taylor, my former officemate, and Heathens. But, as argued, the goods accessible to humans through farming animals cannot justify ignoring animal rights. But maybe there will still be ways to co-live with bees without farming them.[21] And—to jump ahead to Chapter 4—the zoopolis may still contain honey, even without apiculture. So perhaps the result is not *so* bad. Maybe the zoopolis's Taylors still build hives, the zoopolis's officemates still drizzle honey, and the zoopolis's Heathens still toast with mead—all while the zoopolis's bees have rights.

Concluding remarks

I have identified two groups of farmable/consumable animals in the zoopolis. First, animals about whose non-sentience we are (all but) sure are fair game, as plants are. Second, animals who *may* be sentient, but probably not, warrant protection—but this does not take them off the table.

I have proposed that the zoopolis protect these latter animals—Plausibly Sentient animals—with one legal right: the right not to have suffering inflicted upon them. This right protects the animals from the most egregious actual injustices they *would* face if they were sentient, but it leaves open their exploitation, providing those exploiting them take precautions. This marks a viable compromise between the risk of violating these animals' rights and unnecessarily interfering with people's lives.

More than merely *permitting* the farming and consumption of (some) invertebrates, however, I have indicated—drawing upon the arguments developed in Chapter 1—that the zoopolis may well have reasons to *welcome and encourage* their farming and consumption. I have not *proven* that a food system containing these elements is all-things-considered best—but, as explained, proving this would be a mammoth task. But I have offered some key animal-based components of a food system that the zoopolis should be ready to explore.

---

[21] May there still be room for rights-respecting apiculture? In Chapters 5 and 6, I explore labour rights for animals. Could such ideas apply to bees?

3
Plant-based meat

It is no secret that plant-based eating has become more accessible in the west throughout the 2010s and 2020s. One of the developments that has come with increased accessibility—I make no claim about cause and effect—is an advance in availability, variety, familiarity, and quality of plant-based meats.

What are plant-based meats? It is best to first say what they are not. They do not contain slaughter-based meat.[1] They are not simple plant-based proteins, like nuts, beans, or mushrooms,[2] nor manufactured plant-based protein sources that have their own independent culinary histories and purposes, like seitan, tempeh, and tofu. (Though they may contain these.) They are not even the stereotypical (uninspiring) veggie burgers or nut roasts with which long-time vegetarians and vegans are inevitably familiar. (They are, though, these foods' culinary descendants, with lines blurring at the edges.)

We must also distinguish plant-based meats from cultivated meat and the products of cellular agriculture. I explore these in the next two chapters. Like cellular agriculturalists, however, producers of plant-based meats may utilize cutting-edge manufacturing and production methods to create (or imitate) properties or constituents consumers associate with slaughter-based meat. Indeed, plant-based meats may contain the products of cellular agriculture, but I leave this aside.

Plant-based meats are products that culinarily replicate slaughter-based meat as closely as possible that contain only plants.[3] They look, taste, and smell like slaughter-based meat; they have the 'mouthfeel' of slaughter-based meat; they will (if the slaughter-based meat they replicate does) 'bleed'. Producers package them like slaughter-based meat; grocers sell them like slaughter-based meats; cooks serve them like slaughter-based meat. Companies may even manufacture them to look like pieces of a dead animal. For example, the British vegan street-food chain Biff's attracted press attention when they launched vegan chicken wings complete with a sugarcane 'bone' in 2022.

---

[1] Or other meat from the body of animals.
[2] Mushrooms are not plants, of course.
[3] At least, no animal products.

Plant-based meats are familiar to meat-avoiders (vegans, vegetarians, 'reducetarians' ...). Indeed, in the UK, all major supermarkets now stock, and typically even have their own ranges of, plant-based meats. (This was not necessarily true even in the mid-2010s.) The most sophisticated examples of plant-based meats, however, might still have an air of exoticness or exclusivity about them. The Impossible Burger, for example, is not available in Europe. The Impossible Burger famously 'bleeds', and contains haem (or heme), a compound found in blood that Impossible Foods synthesizes from plant-based sources.

To make sense of the idea of plant-based meat, we have to assume a *culinary* metaphysics of meat, according to which a substance is meat if it matches the culinary features of meat. Alternatively, a *historical* metaphysics of meat would demand a substance have a particular history (i.e. a particular relationship to an animal's body) to be meat. A *material* metaphysics would ask what substance constitutes the putative meat. The phrase *plant-based meat* is presumably[4] paradoxical on historical or material metaphysics of meat. But it is perfectly coherent for culinary metaphysics of meat.

I invite sceptical readers to think of bread. A simple bread, of the kind schoolchildren make, might contain wheat, water, and yeast. But we recognize some breads as bread even though their recipes do not contain yeast (i.e. unleavened breads); we recognize some breads as bread even though their recipes do not contain wheat (e.g. cornbreads); and we recognize some breads as bread even though their recipes do not contain water (e.g. milk breads). Thus, the essence of *bread* is not constituent materials or ingredients. We can also separate bread from a given preparation method and still recognize it as bread. For example, bread need not involve baking, as there are raw foods that are still recognizably breads. Thus, bread's essence is not a particular causal history. Just as we can separate bread from particular materials and histories, we can with meat—or so say those who see plant-based meats *as* meats.

Perhaps, then, some advocates of plant-based food systems—namely, plant-based food systems containing plant-based meats—are not advocating for a system in which people lose access to meat. Instead, they are advocating for a system in which we make meat in a different way. Equally, the person who advocates for a food system that (for some reason) does not contain wheat is not advocating for a bread-free food system, as it could still contain non-wheat-based breads.

---

[4] Depending on the specific account.

Let's put metaphysical questions aside. While I use the term 'plant-based meat', this chapter's central question is worth asking regardless of whether these products are meat or not. (Not least as many of the puzzles apply to cultivated meat, which I explore in the next two chapters.) Would the just, animal-rights respecting food system contain plant-based meats? Would we pile barbecues high with meaty (animal-free) burgers, sausages, and steaks in the zoopolis—or would we grill only decidedly less meaty corn, mushrooms, and cauliflowers?

It is my contention that plant-based meats are compatible with animal rights, and that there is every reason to imagine them present in the rights-respecting food system. To make this case, I respond to two sets of critiques. The first comes from animal rights advocates (understood broadly). This is the idea that creating plant-based meat is disrespectful to animals, even if no animals are (directly) harmed in its production. The second comes from the slaughter-based meat industry, food activists, and some vegans. This is that plant-based meats fail as food. It is important to respond to these objections to demonstrate that not only are plant-based meats a *permissible* part of the food system, but they are a *desirable* one, too.

The exploration offered in this chapter takes place against the background of the work done in Chapter 1. That is, we have already seen that communities have good reasons not to ban foods that are not unjust, and, what is more, that they have good reasons to champion meat-based foods if/when they are not unjust. (What precisely is meant by 'championing' will be returned to in Chapter 7.)

For example, plant-based meats may be valuable in helping individuals (foodies; people who work with food; people for whom meats are part of valued cultural practices …) realize their diverse conceptions of the good. If the hotelier's best work is firing up a meaty barbecue on a hot summer evening, if an American family's Thanksgiving just would not be the same without a turkey on the table, if the gourmand's greatest pleasure is foie gras on fresh bread, then all the better if Not Dogs, Tofurkey, and Faux Gras are readily available in the zoopolis, even while slaughter-based meat is not.

Disrespect to animals

In this section, I identify and respond to a range of objections contending that plant-based meats are disrespectful towards animals.

Mixed messages

A vegan might concede that producing plant-based meats is compatible with animal rights, but nonetheless deny that it is permissible to eat them, as doing so sends mixed messages.

Imagine I live with hens rescued from exploitation. These hens occasionally lay eggs, and do not seem to mind my taking them. Perhaps boiling one of these eggs for my lunch would not involve violating any animals' rights (Fischer and Milburn 2019). But imagine colleagues see me eating this egg. Taking a simple (naïve?) view about how my actions impact others, we could imagine whispers. 'Isn't Josh an animal advocate? But he's eating eggs! Perhaps there's nothing wrong with eating eggs.' If, subsequently, each of these (vegan) colleagues goes out and purchases some eggs, *my* rights-respecting eating of *this* egg has resulted in a great deal of *non*-rights-respecting purchase/consumption of eggs.

It is minimally plausible that consuming rights-respecting animal-based foods (in our non-ideal world) could lead others to consume non-rights-respecting food. The trouble is that if the original consumption is rights-respecting, it is difficult to see how the state could prevent it without infringing on *my* rights to freely eat what I like—returning to an earlier idea, this gives the state undue power (Machan 2004, 23). It is not clear that other people's ignorance of our actions is a good reason to stop us from doing something the state should otherwise permit us to do.

To put it another way, this challenge seems to conflate just food systems with activist ethics (cf. Fischer 2020, chapter 9). For animal activists, it may be counter-productive to send messages with eating habits that others could misinterpret. Indeed, it may conflict with moral duties they have *qua* activist.[5] But I do not wish to get into this. It does not concern food systems, and it has little to do with proper behaviour in the zoopolis. There, in contrast to today's world, there would be little chance of mixed messages. My colleagues could not say 'Josh is eating *that* [rights-respecting] egg, so I can eat *this* [non-rights-respecting] egg', as there would not be non-rights-respecting eggs available. Not only would this mean my colleagues could not (easily) violate chicken rights, but they would have no reason to think I was ignoring chicken rights in the first place. (I return to this thought shortly.)

---

[5] There would be ways they could mitigate the risk of misinterpretation—e.g. I could use my egg to provoke conversation about egg-eating.

A mixed-messages objection *might* be a valuable argument against my writing this book. Perhaps, no matter how clear I am about my condemnation of *current* practices of meat-production, by endorsing the production and consumption of plant-based meats, I will encourage some to eat slaughter-based meat. I will put this objection aside, despite its importance. For present purposes, I am interested in putative reasons that it is impermissible to produce and eat plant-based meat, not in putative reasons that it is impermissible to *argue* that it is permissible to produce and eat plant-based meat. I am interested in the first-order questions of justice, not in the second-order question of the ethics of exploring justice.

Finally, there may be a moral wrong in doing something that *appears* to be morally wrong, even though it is not (otherwise) morally wrong. This is the essence of the Jewish law of *marit ayin* (Wurgaft 2020, 154). The critic might note that the consumption of plant-based meat, even if it *would be* morally acceptable, *looks like* (unacceptable) slaughter-based meat consumption. Therefore, consuming plant-based meat is unacceptable.

This is a moral argument, but perhaps we could offer a parallel political argument. Maybe the state may restrict that which *appears* unjust to discourage unjust behaviour and/or assure citizens. Perhaps, for instance, the state may require violent theatrical performances to take place behind closed doors, and not on street corners. Analogously, perhaps the state may ban the consumption of plant-based meat, even though it is not unjust, to avoid sending the wrong messages.

But neither argument (moral nor political) is compelling in the current case. In the zoopolis, a citizen's consumption of plant-based meat would not appear to be wrong. There would not be slaughter-based meat available, and so onlookers would not mistake the consumption of plant-based meat for the consumption of slaughter-based meat. Equally, today, no one would mistake the consumption of beef for the consumption of panda—even if a casual observer might be hard-pressed to differentiate beef from panda meat. Someone who believed it wrong to consume panda would be confused in condemning a beef-eater in a Berlin restaurant for doing something that *looked like* eating panda meat.

Representation and use

Let us explore a different worry. In producing plant-based meat, perhaps 'we are participating in the nonartistic representation of nonhuman animals as mere resources' (Turner 2005, 6), and this, in turn, violates animals' rights: 'if nonhuman animals have the same basic rights human ones do, then all nonhuman animals have a basic right to autonomy and so not to be represented as a mere resource' (Turner 2005, 4–5).

There are several problems with this argument. One is that we should not simply assume that 'animals have the same basic rights human ones do'—even if we 'avoid the more immoderate version of the animal rights position' (Turner 2005, 4–5). Instead, we should ask whether animals have a sufficiently strong *interest* in the same things that humans have (putative) rights to. For example, do animals have an interest in autonomy? In not being 'represented as a mere resource'? The claims require evidencing.

But imagine that animals do have such interests. Still, the creation of plant-based meats need not involve representing animals as mere resources. We can see this by considering the ways in which we replicate parts of human's bodies, or products from their bodies.

We make infant formula approximating breastmilk, while we lull babies to sleep with sounds imitating those in the womb. We create and fit artificial organs (hearts, lungs …) or other body parts (limbs, joints …) for people with disabilities and diseases. We replicate human hair for wigs, whether for fashion, art, celebration, or to assist those who have lost hair. We replicate human teeth with caps and implants for those who have lost or damaged their own. To train medical and cosmetic professionals, we replicate hands, mouths, reproductive organs, and more. We replicate human bones for educational purposes, Hallowe'en decorations, or film props.

This is just a taster of the ways that we imitate parts or products of human bodies. These imitations can be controversial, and we should not pretend that they are entirely ethically innocuous. But it would be odd to hold that all these imitations were (prima facie or all-things-considered) rights-violating for representing humans as mere resources.

This shows us something important. Even if someone has an interest in autonomy and an interest in others not viewing them as a mere resource—as (many) humans surely do—we can sometimes replicate parts or products of their bodies while respecting these rights. Thus, the idea that there is something *inherently* rights-violating about the creation of plant-based meat because it resembles part of an animal's body is a non-starter.

It is not the case that plant-based meat is rights-violating for approximating something that comes from someone else's body. Instead, we should ask what kinds of protections, if any, we should implement so that we do not harm the being whose bodies we 'represent'. It is not clear how we harm animals by producing plant-based meat. It does not harm them directly, of course. But perhaps there are indirect harms. There are several possible candidates for such harms, to which I now turn.

Meat as food: Society

For some vegan theorists, plant-based meat serves to reaffirm meat as food, and, for this reason, we should reject it. We can distinguish two different forms of this argument: in one, the objection is to our reaffirmation of the place of meat in society. In the other, the focus is more personal. I address the latter shortly. For now, I focus on the argument that plant-based meat affirms a problematic vision of meat *in society* (see, e.g., Alvaro 2020, 84–85; Sinclair 2016).[6]

The argument is, roughly, as follows. In a given sociocultural context, meat has a certain centrality: (particular) human communities are *carnophallogocentric*, or *carnist*, or have a *meat culture*.[7] Not only this, but meat has meanings concerning the superiority of certain ways of being. The most obvious of these is *human*; the culture links meat to human superiority. But other analyses tie meat consumption to interhuman hierarchical relationships. Scholars identify connections between meat and the value of being male, white, language-using, able-bodied, and more. The centrality of meat is a worry, these scholars argue, not just because the (slaughter-based) meat industry harms animals, but because of meat's problematic meanings. Plant-based meat might well be able to help animals harmed in the production of slaughter-based meat, but it does not challenge the place of meat in society. Instead, it *reaffirms* it.

I do not want to challenge these analyses about meat's cultural meanings. But I do want to identify two limitations in using the analyses to challenge plant-based meat.

First, arguments challenging the consumption of plant-based meat because it reaffirms problematic visions of the place of meat in society risk stressing some associations of meat over others. Yes, perhaps a given culture traditionally ties meat to the superiority of (certain) humans. But meat also has extremely *positive* (varying) cultural associations. Many associate meat with

---

[6] See Miller 2012 and Cole and Morgan 2013 for similar criticism of cultivated meat.
[7] Jacques Derrida's *carnophallogocentrism* (Adams and Calarco 2016) is the thought that the subject (thus the rights bearer) presupposed in western thought is language-using (*logo-*), virile/masculine (*phallo-*), and meat-eating (*carno-*). Melanie Joy's *carnism* (2009) refers to the (often hidden) ideology that frames some animals as edible. *Meat culture* is 'all the tangible and practical forms through which the ideology [of carnism] is expressed and lived' (Potts 2016, 19).

good health: it is nutritionally rich and satisfying to eat. Many associate meat with prosperity: it is good-quality food, frequently the centrepiece of meals. (A 'chicken for every pot' means we are *all doing well*.) And, as explored in Chapter 1, meat can be an important part of many practices of religious, cultural, and personal significance: it isn't Christmas dinner without turkey, it isn't watching the Superbowl with dad without chicken wings, it isn't Passover without chicken soup.

Second, these cultural arguments must not overlook how associations and cultures can change and evolve. This is certainly true of food. Quickly shifting cultural norms can see assumptions about what constitutes food—never mind the associations of particular foods—completely change in the space of a generation or two. For example, whale meat was an important part of the British diet in the Second World War and post-war years. Over the following decades, British involvement in whaling plummeted, and the import of whale meat ceased. I have never eaten whale meat, never truly thought of whale meat as food, and never knowingly been in the presence of whale meat. On the other hand, my grandparents ate it when they were children.

Or let us consider an example of food *arriving*. A standard example of changing food cultures is the sudden acceptance of sushi in North America, but something similar happened in Britain. Although by 2011, an article in *The Independent* could (provocatively) proclaim sushi 'Britain's new national dish' (Walsh 2011), the first Japanese restaurant did not open in Britain until 1974, while sushi only started becoming popular among certain groups (i.e. wealthy London businesspeople) in the 1980s. Today, sushi bars are a familiar part of the British high-street.

Just as whale meat was a part of my grandparents' generation but alien to my own, I often eat sushi, but suspect that none of my grandparents ever ate it. Indeed, they may not have known what sushi was, or even recognized it as food. If what counts as food can change so quickly, it should not be surprising that associations and the cultural roles of foods can change quickly, too. Real ale changed from something for older, rural men to something beloved by educated urban youngsters; kale moved from a winter fodder crop to a superfood ubiquitous among the health-conscious; desserts considered extravagant by my parents' generation look to me, at best, kitsch.

And, crucially, changes in production methods can revolutionize what foods mean. Refrigeration, increased farming efficiency, canning, and 'TV dinners' all, in their own way, revolutionized what certain foods could be, what they needed, what they *meant*. Foods for the rich became accessible, losing their sophisticated air. Foods requiring less preparation allowed women—the cooks—to pursue their lives, challenging what it means to prepare food (and to be a mother, wife, woman). Seasonal foods became accessible year-round, impacting associations between food and times. And globalization allows the same food to mean something very different in very different contexts. In the American South, complex racial politics entwine fried chicken. But in urban Japan, it is a (surprising) Christmas tradition. And *plant-based* chicken complicates both stories.

The societal arguments against plant-based meat risk missing the positive associations of meat, and the way associations of foods can change (sometimes quickly). And even if the overall associations of meat are negative—which is not clear—that need not be a reason to use the state's coercive power to *ban* plant-based meat.

Instead, it might be a reason to seek to *change* associations. Producers and consumers of high-quality ciders challenge associations of cider with alcoholism; the ubiquity of wines in supermarkets challenge associations of wines with class superiority; and indigenous communities reclaim appropriated foods. It would be surprising to say the state should ban cider (including fine ciders) because of White Lightning,[8] ban wines (including affordable wines) because of oenophilic snobs, or ban quinoa (including subsistence consumption in the Andes) because of Instagram. Surely, the answer is to draw attention to alternative associations of these foods, and champion decent cider, affordable wines, and the indigeneity of quinoa.

Something similar, I suggest, is true of meat. Might there be a way to challenge the negative associations of meat but champion its positives? Might we change the negative associations of meat, just as associations of other foods (including specific meats) have changed in the—very recent—past, and will continue to change?

I contend that plant-based meat provides an opportunity to separate meat from the killing of animals. Through a shift from slaughter-based to plant-based meat, meat could stop being associated with relations of domination over animals. Indeed, meat could become associated with *positive* co-relations with animals; we can build new cultural associations in response to the changing place of animals in our food system. (Compare Chapter 5.)

Were we to sever the association of meat with human superiority, the cultural place of meat in relation to oppressed human groups, too, would shift. No longer would meat be about hierarchy, domination, and death. Central to the image of meat as championing certain white, masculine, language-using (etc.) perspectives is an association of animals with non-white, feminine,

---

[8] White Lightning was a brand of strong, cheap cider available in the United Kingdom in the 1990s and 2000s that became a byword for alcohol abuse.

dumb (etc.) people. But if meat no longer involves the denigration of animals, then it no longer involves the denigration of these people (cf. Dunayer 1995).

Indeed, Carol Adams, who provides the touch-stone conceptualization of meat's meanings for vegan cultural/critical theory in her *Sexual Politics of Meat* (1990), offers similar arguments. Central to her analysis is the animal as the 'absent referent'—in meat, she argues, the animal is absent. Women, too, can be absent referents when a patriarchal visual culture presents their bodies for consumption. That women and animals share in being the absent referent is central to their co-oppression, and the links between patriarchy and speciesism. Thus, Adams links feminism and veganism.

Plant-based meats, however, can represent animals quite differently: a pig pictured on the packaging of plant-based pork products (Adams argues) may 'bespeak a presence rather than an absence'; may be 'a representation of those bodies allowed to live' (Adams 2016, 252–253). This presence reverses meat's negative associations.[9] Alternatively, plant-based meats could sever the link between animals and meat altogether. Equally, a generation raised with camera phones ('phones') and digital cameras ('cameras'), does not associate photography with film (Friedrich 2020, 23).

Wholefood, plant-based eating—including a diet containing the 'meaty' seitan that Adams (2016) takes pride in preparing—could provide an important site of resistance to the dominant sexist, meat-eating culture. But so could replacing slaughter-based meat with plant-based meat, especially if we rethink our relationship to animals. More, a *food system* that replaces slaughter-based meat with plant-based meat, especially when combined with a rethinking of the place of animals (and oppressed humans) in the food system, could provide an important part of the alternative to the dominant sexist, meat-eating culture.

Meat as food: Individuals

Let us turn to the more individual-focused argument about reaffirming meat as food. These arguments do not look to the cultures around meat-eating, but what the eating of meat (plant-based or otherwise) says about individuals. The thought is that, although plant-based meats are *preferable* to slaughter-based meats, there is something morally dubious about consuming them, due to the attitudes they express towards animals.

---

[9] Plant-based meat does not kill, or make-absent, anyone. The plant-based steak contains plants, which are not someones. Thus, for Adams, we have no reason to be worried about plant-based meats for their own sakes.

Imagine, for example, a detective investigating a modern Ed Gein. Gein, an inspiration for *The Texas Chain Saw Massacre*, was a serial killer and grave robber noted for making trophies out of human bodies. Let us imagine this detective spots a human-skin lampshade in Gein's house, which he admires so much that he commissions an exact replica of it, made from a plant-based human-skin analogue. There is something intuitively troubling about the detective's actions—at least, according to Bob Fischer and Burkay Ozturk (2017). Analogously, there is something troubling about the actions of the person who respects animal rights but nonetheless seeks out imitations of animal flesh.

It is possible that our intuitions are awry in the detective case. Perhaps, for example, they are off-kilter because of our disgust at the thought of a serial killer skinning his victim and then crafting things with their skin. If so, perhaps we should accept that the detective (who is not killing or skinning anyone) is not doing anything wrong.

I have some sympathy with this view. There is nothing immoral about engaging with the macabre—there is nothing wrong with enjoying horror literature, films, and video games; with reading biographies of and watching documentaries about serial killers; with decorating our houses for Hallowe'en. I play horror-themed live-action role-playing games; I have a friend who records podcasts about disasters resulting in catastrophic loss of life; I have a cousin who performs a comedy routine about her collection of creepy dolls. It takes all sorts to make a world. Maybe the detective's interests are *too* macabre for some, but condemning on those grounds sounds like prudery.

However, we need not rely on intuition to identify the immorality of the detective's actions (Fischer and Ozturk 2017). Virtue ethicists might claim that the virtue of *reverence* calls for us not to admire a human-skin lampshade. Alternatively, those sympathetic to A. E. Moore's ethics might say that admiring a human-skin lampshade involves admiring things that are bad. By analogy, the virtue ethicist or Moorean might reject plant-based meat.

I do not want to get too far into these arguments. Both are distinctively *moral* arguments, and not presented to justify banning plant-based meat (or otherwise removing it from our food system). That said, perhaps we could use them to argue for a ban. While such arguments hold limited sway among liberals, some support banning the deeply immoral *because* it is immoral. Patrick Devlin's (1965) arguments supporting the criminalization of homosexual sex and prostitution are a stark example. Other examples might include laws about drug-use and gambling; public sex/nudity, swearing, or drunkenness; access to violent, disturbing, or explicit media; and so on. (I do not want to oversimplify; there are many arguments for these laws.

My point is that laws based on appeals to citizens' moral sympathies might not be *so* alien.)

I suspect that there are problems with these approaches when it comes to opposing the consumption of plant-based meat—whether as a 'merely' moral matter or using the law's coercive power. There are disanalogies between the detective and the eater of plant-based meat, and that comes from the way that the detective first encountered the lampshade. Humans' encounters with meat, however—especially in the zoopolis!—are going to be quite different.

Perhaps we could concede that there is something unvirtuous (or an admiration of that which is bad) in visiting a slaughterhouse, watching slaughterpersons kill and butcher an animal, and thinking 'That looks great. Maybe I can get something like it.' On the other hand, there is nothing obviously unvirtuous (and no obvious admiration of the bad) in a shopper thinking 'Those plant-based prawns look *just* like the ones I had in paella on holiday when I was a kid.' Once again, meat is not *all* bad. Even if we must condemn the violation of animals' rights, including in food production, not everything associated with animal foods is evil. There are thus, I suggest, ways of desiring meat that are neither unvirtuous, nor an admiration of that which is bad. It comes down to *why* something is desired.

As explored above, plant-based meat gives us the opportunity to *remove* those bad things from meat production. It gives us a chance to change the *meaning* of meat. If we no longer associated meat with violence, it is hard to see why desiring meat would entail any attitude towards harm to animals. Compare: It is not, today, unvirtuous or an admiration of that which is bad to desire freshly made bread, even though farmers once violated the rights of animals (e.g. oxen) to produce grains. Equally, in the zoopolis—where no one abuses animals for meat—it would not be unvirtuous, or an admiration of that which is bad, to desire meat just because, *at one time*, agriculturalists produced meat disrespectfully.

Symbolic disrespect

Developing the previous critiques, we could argue that it is disrespectful to eat meat—or engage in institutions surrounding meat-eating—because of what it symbolizes. We could say that plant-based meats are a celebration of harmful practices, and this is disrespectful to those animals who were, in the past, killed or otherwise harmed in the pursuit of meat.

In the previous subsection, I explored the idea that an individual's consumption of plant-based meat might not show appropriate reverence to animals killed. Perhaps, similarly, a zoopolis with plant-based meat is a *society* insufficiently reverential to those animals on whose graves the society stands (compare Scotton 2017).

I illustrate this by way of an example.[10] Imagine an English village has an annual witch-burning fête, in which villagers parade a mannequin of an elderly woman through the street and then burn it. At first glance, this might seem eccentric, but little more so than many folk traditions. But we might nevertheless think that the fête is problematic in that it fails to properly respect the memory of the thousands of people, mostly women, executed as witches in Britain. Authorities tortured, humiliated, and then (typically) hanged these innocents. This is not the sort of thing we should be making light of—this is not the sort of thing we should be celebrating, or uncritically re-enacting. It is one thing for us to re-enact a witch trial as an educational exercise or as tribute to, or remembrance of, the dead. It is quite another for us to make nameless victims of witch trials a mere prop.

Perhaps some of the practices associated with plant-based meat are comparable to the witch-burning fête. For example, we might feel there is something distasteful about people appropriating butcher imagery to sell plant-based meat.[11] We could argue that dressing up as a butcher often involves disrespect to butchers' past victims—or, more broadly, that running through rituals adjacent to animal harm involves a making-light of that harm, as with the witch-burning fête. Although—on this argument—no present animal has reason to complain, perhaps we have duties to respect the memory of the dead. Indeed, we could make a similar argument about plant-based meats themselves,[12] and not just dressing up as butchers to sell them.

How can we respond to this? We could deny that we have the kinds of duties of remembrance that the argument posits. Perhaps, for example, this is a moral matter, and so not the sort of thing that the state should be involved with. If some people choose to respect the memory of the dead by condemning activities and rituals that evoke harm, so be it. If others are not worried about this issue, then that is their prerogative. But we may still want to hold that the state should prevent people from engaging in re-enactments of witch-burning.

---

[10] There are contentious examples associated with continuing racism and colonialism, as well as sophisticated academic literatures around them. Animal activists face (sometimes fair) criticism for clumsily deploying comparisons between animal harms and interhuman abuse. I have no desire to court controversy or offence, and so the case I use is less familiar—although not, I hope, less illustrative.

[11] In the zoopolis, specifically. There are a range of *non-ideal* considerations that may speak strongly in favour of butcher imagery used to sell plant-based meat in today's world.

[12] We should distinguish this argument about respect for the dead from a separate argument about respect for corpses. While it is wrong to disrespect animal corpses (Milburn 2020), plant-based meats are not corpses.

We could make such an argument by claiming that (some of) the wrong in the witch-burning case comes from its disrespect towards *actually existing* people. There are a few ways to do this.

One deploys an indirect duty argument. Some existing people care about the women killed as witches, and we have duties to *these existing people*, providing prima facie reason to appropriately respect the women killed in earlier centuries. (Equally, we could reasonably describe the burning of a flag as a prima facie wrong towards patriots.) But while this does offer us (defeasible) moral reasons to change our behaviour—the fact you consider it important to respect the flag gives me some reason not to burn it—it does not provide a compelling reason to ban practices.

For the most part, the coercive power of the state should not restrict *me* from engaging in practices merely because *you* find them distasteful. This is especially so when the supposedly distasteful conduct is important to me (as meat-eating may be). Were this not so, we would quickly reach a situation of state repression of religion. Christian practices may well be offensive or distasteful to some Muslims, whose practices may, in turn, be offensive or distasteful to some Christians. For example, was Jesus divine? ('How *dare* you say the Christ was a mere man?' 'How *dare* you equate a mere man to Allah?') Even if some animal activists do deem the creation or consumption of plant-based meats disrespectful or offensive, this is not a good justification for state action.

A more promising route to arguing that witch-burnings are harmful does not look to existing people *who care for* those killed, but existing people *like* those killed. Perhaps the opponent of witch-burning could look to elderly women, Wiccans, or the people who continue to face persecution as witches. After all, perhaps the fête threatens or disrespects these people *themselves*, justifying the use of state power. Equally, perhaps dressing as butchers involves disrespect towards existing animals.

There is an easy response. Animals will not feel threatened or disrespected by 'butchers'. Older women, Wiccans, and others may feel threatened by 'witch burners'. This difference seems relevant.

But this is too quick. The harm to (say) Wiccans by witch-burnings may not be a matter of the subjective states evoked by the practice, and so the fact that animals do not have those experiences may not matter. I have elsewhere argued that animals could warrant protection across many theories justifying legal proscription on hate speech, *even though* hate speech does not (directly) evoke aversive subjective experiences (e.g. fear) in animals (Milburn and Cochrane 2021). So, for example, perhaps the state justifiably censures hate speech because hate speech creates a hostile environment. According to this argument, hateful utterances are not *themselves* harmful, but they lead to an environment in which members of targeted groups are more likely to face violence. Thus, hate speech contributes to harm, meaning the state can deploy coercive power to prevent it.

I am not saying this is a fool-proof argument for anti-hate-speech laws. I am saying that this argument (among others) in favour of legislation applies equally well, in principle, to hate speech targeting both humans and animals.

This observation is important for the case against plant-based meat that we are exploring, given that hate speech is not solely about *speech*. We might categorize all kinds of symbols (activities, dress, rituals ...) as hate speech. Perhaps one does not need to *say* anything to engage in hate speech against Wiccans or pigs—perhaps simply dressing up as a witchfinder or butcher (in the 'right' context) is enough. And if that is so, it may sometimes be appropriate to deploy the state's coercive power to stop people dressing as witchfinders or butchers. (Although I remain with hate speech for this subsection, we could express the arguments in other terms.)

What should we make of this argument? Let us assume, first, that the zoopolis should have laws against hate speech.[13] This granted, it seems right to say that, *in principle*, re-enacting certain symbolic rituals (or wearing certain symbolic clothes, or similar) associated with meat could constitute hate speech against animals. But, here, it is helpful to again turn to the distinction between *why* certain food-related practices are desired or performed.

Some motivations are clearly hateful, and some are not. On the face of it, for instance, there is a difference between dressing as a butcher because you wish to affirm 'butcher ideology' and dressing as a butcher because this is the uniform of your decades-old family business, of which you are proud. (Why be proud of a butchery? Perhaps I would be proud of my family business if it happened to be one of the first *plant-based* butchers.) Similarly, there is a difference between desiring a 'bleeding' burger because the 'blood' has culinary benefits—improved taste, improved texture, a nostalgic sizzle—and desiring a bleeding burger because it helps you pretend that you are supporting slaughter. Just as (I argued above) *some* reasons for desiring plant-based meats may be virtuous, so some symbols associated with meat-eating may not be hateful.

I do not want to take a stance on *which* symbols are hateful, as this would depend upon which theory of hate speech the zoopolis endorsed (if any), and (perhaps) empirical factors. But I suspect that examples are few and far between. Pertinently, I am not of the view that producing or eating

---

[13] I am not committed to that claim.

plant-based meat is hateful. To see why, it is worth returning to my earlier comments about the cultural significance of meat. While meat can indeed have cultural associations with human (male, white …) superiority, it can also have deeply *positive* associations.[14] It is overwhelmingly favoured/desired by people because of its positive associations. (This includes the valuing discussed in Chapter 1.) It is, for the people who desire it, good food—tasty, healthy, nostalgic, convenient, culinarily versatile, culturally significant, spiritual, and more. But none of *these* reasons for eating meat are hateful, and people eating plant-based meats for *these* reasons should not, I think, face censure for engaging in symbolically hateful activities.

But this cannot be the end of the matter. Some things are hateful *regardless* of intention. Maybe our witch-burning villagers are engaging in hate *even if* the burning—for them—is about childhood memories of building bonfires. Equally, certain rituals associated with meat-eating could be symbolically hateful *even if* they are carried out for the most life-affirming reasons.

It is worth returning to the chance of changing the meanings of meat in the zoopolis. Plant-based meats give the opportunity for meat to mean something different. Even if—a claim to which I am unsympathetic—meat-eating is *today* hateful regardless of intentionality, that is not to say that it will continue to be so in the zoopolis. Meat can come to mean something different. Indeed, it could even come to mean something *pro*-animal.

Symbols are not hateful *inherently*, but because of the meaning that we give to them. And those meanings change. The fact (if a fact) that Hengist and Horsa led Germanic invasions into Celtic Britain, and the fact that a horse's head was their symbol, does not mean that to wear horse-head jewellery is to engage in Celtophobic hate speech. The idea is almost absurd. If cultural associations of meat-eating are today hateful, then I contend that we should be trying to change that. The first step will be severing the link between meat and harm—which is precisely what plant-based meats offers.

Bad food

We can now turn to a range of critiques of plant-based meats that challenge them not for animals' sake, but because they are, for some reason, poor foods.

---

[14] Not for animals, maybe. Or not for *some* animals. But it is inevitable that associations between foods vary between different subcommunities in a society. Generational, religious, class, geographic, and—we can add—*species* groups within the same community are going to associate the same food with different things.

Against meat

First, we can respond to those vegan critics who object to plant-based meat as unnecessary *because it is meat*. Some vegans—like those who make the various meat-as-food objections discussed above—argue that while plant-based meat does not involve harm to animals, we should reject plant-based meat because meat is not *good* food: it's disgusting, it's unnecessary, it's unpleasant, it's unhealthy. Why create plant-based meats, these people ask—why not adopt, say, raw veganism (Alvaro 2020)? Would it not be *unvirtuous—intemperate*—to favour plant-based meat when wholesome plant-based diets are an option?[15]

There are at least four responses to this. The first is a decidedly non-ideal answer, and so not my focus (cf. Milburn 2016). Perhaps it would be good if everyone immediately switched to a whole-food plant-based diet. But only some people do, so it is better to provide plant-based meat for those who do not make the leap.

Second, as detailed in Chapter 1, there are many good things about meat, and many things we would lose if we did away with meat. Even if some people find meat disgusting or unpleasant, some people find it delicious. Even if some people find meat unnecessary, some people find that it is an important part of their diets, cultural expression, personal relationships, and more. That is, many of the people who value meat are not simply 'overestimating the importance of taste' (Alvaro 2020, 138). (And, echoing Chapter 1, we should not *under*estimate taste's importance. The pursuit of good foods can be an important component of the good life.)

While meat *can* be unhealthy, we would struggle to find a qualified dietician who denied that we can healthfully enjoy small amounts of meat (or some meats). Indeed, meat can be an important source of nutrition for food insecure people, or people with certain health conditions. (I return to this thought in Chapter 5.) So, there are good reasons not to simply embrace a raw vegan food system—assuming that there are rights-respecting alternatives to a raw vegan food system.

Third, we should be reluctant to say that these are the sorts of reasons that justify state use of coercive power. One need not be a liberal or libertarian to believe that the state should not prevent *me* from doing something merely because *you* find it disgusting, especially when that thing is important to me. (*Perhaps* the state might mandate that I do it behind closed doors.) It

---

[15] Carlo Alvaro (2019) offers this final critique to cultivated meat, but his arguments apply to plant-based meat—he groups both as 'mock meat' (Alvaro 2020, 51). Elsewhere, he criticizes plant-based meat for replicating meat, which, he reports, he finds disgusting (2020, 85).

is also hard to imagine that the state should be determining some foods *so* unpleasant that no one may eat them. That is absurd—many *do not* find them unpleasant. And even if we could find a dietician who insisted that the only heathy amount of meat is no meat at all, it is not the role of the state to force us to eat healthily (cf. Barnhill and Bonotti 2022). Some of us value access to foods we enjoy, consider culturally or personally valuable, or similar more than we value a completely healthy diet. And that is fine. Others value the pursuit of healthfulness—that is fine, too. Neither should impose a particular understanding of the good on the other.[16]

Fourth, there are reasons to think plant-based meat might be *less* disgusting, unpleasant, or unhealthy than slaughter-based meat. Disgust is, I think, appropriately directed towards slaughter-based meat because of what butchers and farmers do to animals—but plant-based meat's production is innocuous. Equally, perhaps some chocolate is disgusting because enslaved people picked the cocoa. But there is no need for disgust to linger over all chocolate produced for evermore.[17] Meanwhile, plant-based meat creates opportunities for the development of healthy, safe meat. (More on this later.)

Some people in the zoopolis may believe meat disgusting, unhealthy, and all the rest. They would be free to avoid it. And they would be free to encourage others to abstain. But their beliefs, *even if true*, would not be good reasons for the state to ban plant-based meat.

Against processed food

Although it is less frequently discussed in academic work, a common challenge to plant-based meats in the popular press is that they are ultra-processed, and therefore consumers should be wary of them (see, e.g., Blythman 2018; O'Connor 2019).

This challenge generally involves a comparison between plant-based meats and slaughter-based meats. I rule out slaughter-based meats as unjust, so I will only nod to these considerations. But advocates of a wholefoods plant-based food system also make the challenge, arguing that the food system should contain protein with beans, nuts, tofu, tempeh, and similar, but not plant-based meats. Thus, the issue is worth exploring.

---

[16] There is a line that individuals can cross—especially in states with taxpayer-funded healthcare. States could legitimately take *some* steps to preserve health. I doubt that banning plant-based meat would be a proportionate and effective step towards this goal, though.

[17] Unsurprisingly, aesthetic arguments against meat typically focus on animal agriculture, and especially slaughterhouses, rather than meat itself. See, e.g., Alvaro 2020, 29–34; Holdier 2016.

NOVA[18] is a four-group classification of foods by level of processing (Monteiro et al. 2017, 9–10):

1. Group one are 'unprocessed or minimally processed foods'. Thus, I split it into two:
   a. Unprocessed foods are edible parts nature separated from nature.
   b. Minimally processed foods 'are natural foods altered by processes that include removal of inedible or unwanted parts, and drying, crushing, grinding, fractioning, filtering, roasting, boiling, non-alcoholic fermentation, pasteurization, refrigeration, chilling, freezing, placing in containers and vacuum-packaging'. These processing methods aid in the preservation of natural foods, or make them safer or more palatable.
2. Group two are 'processed culinary ingredients'. These are 'are substances derived from [group one] foods or from nature by processes that include pressing, refining, grinding, milling and drying'. These processes create products that are not typically meant to be consumed on their own, but are used in food preparation.
3. Group three are 'processed foods'. They have undergone one or more 'preservation and cooking methods'. For the most part, they have only two or three ingredients, and remain recognizable as versions of group one foods.
4. Group four are 'ultra-processed foods'. These contain little, if any, intact group one food, instead being made up of 'substances derived from foods and additives'. They 'also include other sources of energy and nutrients not normally used in culinary preparations'. Additives *only* found in ultra-processed foods include 'dyes and other colours, colour stabilizers; flavours, flavour enhancers, non-sugar sweeteners; and processing aids such as carbonating, firming, bulking and anti-bulking, de-foaming, anti-caking and glazing agents, emulsifiers, sequestrants and humectants'.

Plant-based protein sources fall across all four groups. Beans will typically be in group one. High-protein flours will typically be in group two. Tofu and tempeh will typically be in group three. Plant-based meats, meanwhile, will (almost) invariably be in group four.

What should be clear from the NOVA classification is that food processing is familiar. Certain raw diets may avoid most processed foods, but a

---

[18] *NOVA* isn't an acronym; it is just a name.

diet without processed foods is atypical. Indeed, *ultra*-processed foods are familiar. The following are ultra-processed:

> carbonated drinks; sweet or savoury packaged snacks; ice cream, chocolate, candies (confectionery); mass-produced packaged breads, buns, cookies (biscuits), pastries, cakes and cake mixes; breakfast 'cereals', 'cereal' and 'energy' bars; margarines and spreads; processed cheese; 'energy' drinks; sugared milk drinks, sugared 'fruit' yoghurts and 'fruit' drinks; sugared cocoa drinks; meat and chicken [sic] extracts and 'instant' sauces; infant formulas, follow-on milks and other baby products (which may include expensive ingredients); 'health' and 'slimming' products such as powdered or 'fortified' meal and dish substitutes; and many ready-to-heat products including pre-prepared pies and pasta and pizza dishes; poultry and fish 'nuggets' and 'sticks'; sausages, burgers, hot dogs and other reconstituted meat products; and powdered and packaged 'instant' soups, noodles and desserts.
>
> (Monteiro et al. 2017, 17)

Thus, the challenge to ultra-processed foods is certainly not unique to plant-based meats.

Why oppose processed foods? Scientists proposed the NOVA classification scheme because of the health impacts of ultra-processed foods, especially the population-level impacts of food systems incorporating high levels of ultra-processed foods.[19] At the population level, high consumption of ultra-processed foods leads to negative health outcomes including obesity, diabetes, dyslipidaemias (like high cholesterol), and hypertension (Monteiro et al. 2017, 10). This is because ultra-processed foods are typically high in fat, sugar, salt, and calories. But this is not necessarily the case.

Processing itself is not unhealthy,[20] and is near-ubiquitous. Even raw foodists may eat—for example—fermented foods. What is more, processing is necessary to make many perfectly healthful foods edible. Potatoes, aubergines (eggplants), and some beans are inedible when raw. Ultimately, it is 'a mistake to make any judgement of food supplies or foods simply because they are "processed". ... Verdicts on food processing as such have little or no meaning' (Monteiro et al. 2017, 9). Consequently, we must consider the healthfulness of processed foods on a case-by-case basis; it would be

---

[19] Questions about ultra-processed foods at the population level also raise food-justice concerns, insofar as ultra-processed foods can represent (e.g.) multinational corporations displacing established food networks.

[20] Raw foodists sometimes hold that most or all food processing is necessarily unhealthful, but their position is, at best, a minority one.

a mistake to label plant-based meats unhealthy merely because they are ultra-processed. There could be plant-based meats—now or in the future—that are not particularly unhealthy, even if ultra-processed.

Generally, plant-based meats offer roughly the same nutritional profile as the slaughter-based meat products they emulate (Bohrer 2019). That said, plant-based meats do bring with them *other* health benefits relative to slaughter-based meats. (Without the need for farms, for example, there is lower zoonotic risk and antibiotic use.) But it remains unclear whether plant-based meats can replicate the health benefits associated with diets rich in plant-based wholefoods like legumes—even while these are frequently the primary ingredients of plant-based meats (Santo et al. 2020, 8). There is every reason to believe that a plant-based wholefood system would be, overall, healthier than a plant-based food system incorporating plant-based meat.

Might this provide an argument against including plant-based meats in the food system of the zoopolis? As above, we should not endorse *banning* foods simply because there are more healthy options. Individuals might legitimately sacrifice a little health for a gain in some other areas (e.g. a shared experience with friends). In any case, otherwise healthy lifestyles can healthfully incorporate even junk food in moderation. But perhaps the zoopolis would have reason to *favour* healthier protein sources when it must make judgements. For example, perhaps—even if they belong in supermarkets—*schools* should not be serving plant-based meats. (For more, see Chapter 7.) At least, perhaps states should favour a food system containing healthier rather than less healthy plant-based meats. But this is not a simple matter. As seen in Chapter 1, we have reasons to favour a food system containing meat—and thus there may be reasons for the state to support (not merely permit) a food system containing plant-based meats, *even if* a food system without them might lead to healthier diets.

Another worry about processed foods concerns environmental impact.[21] Processing requires energy, and processed foods require packaging. Unsurprisingly, this is not a compelling argument in favour of slaughter-based meat. While 'processed food is more environmentally damaging' may—*may*—be a useful heuristic, a much *more* useful heuristic is that slaughter-based meat products are more environmentally damaging than plant products (Lestar 2021).

---

[21] Alvaro starts and ends his case for raw veganism in these terms. Cooking (a form of processing), he tells us, 'requires the use of ovens, stovetops, microwave ovens, which use coal, wood, gas, and electricity. Not to mention that production of food in factories also requires energy consumption' (Alvaro 2020, 3). A raw vegan diet, however, 'would drastically reduce the use of energy resources' (2020, 128).

We can see this starkly with side-by-side comparisons. A life-cycle assessment comparing the Beyond Burger (a popular plant-based meat burger) to a slaughter-based beef burger, for example, found that the Beyond Burger produced 90 per cent lower greenhouse-gas emissions and required 46 per cent less energy, >99 per cent less water, and 93 per cent less land (Heller and Keoleian 2018). Indeed, examination has repeatedly demonstrated the environmental credentials of plant-based meat relative to slaughter-based meat (e.g. Poore and Nemecek 2018; Santo et al. 2020). As Matthew Hayek and Jan Dutkiewicz (2021) write, counter-claims 'appear to be more rooted in broad opposition to food technology rather than a true environmental accounting'.

But, as animal rights rule out slaughter-based meat, this is by-the-by. All else equal, it is *probably* the case that we need more energy to turn beans into plant-based meats (and then package, freeze, distribute, and cook said plant-based meats) than to pick beans from the garden and serve them. This is a complicated matter. Economies of scale may not favour home growing, and one advantage of processing is that it makes palatable/desirable foods that might otherwise go to waste. (Beans are a telling example, as we generally do not eat them raw. Broad beans, for example, likely go through multiple stages of processing: (double) podding; cooking; refrigeration …) But the point stands. 'Taking a broad view of sustainability', concede Hayek and Dutkiewicz (2021), 'the clear winner is a diet based on whole plant foods—just vegetables, grains, fruits, and legumes'.

We must consider this disparity when making decisions about food systems. But we must balance it against other concerns, including the strong presumption against banning non-rights-violating practices. Mirroring arguments about health concerns, I suspect that the (possible) environmental negatives of plant-based meats relative to plant-based wholefoods do not give us compelling reasons to ban plant-based meats. But they could provide *a* reason that the state should favour a food system rich in plant-based wholefoods, rather than plant-based meats.

This reason, however, need not be decisive. To repeat, there are many ways a food system with (plant-based) meat may be preferable to a food system containing only plant-based wholefoods. These reasons may justify states favouring food systems that are environmentally less-than-ideal.

For example, if the plant-based meat industry produces good, meaningful work, or if plant-based meats are an important part of practices considered central to people's identities or relationships, perhaps the state should support a food system containing plant-based meats even if it is not the most environmentally friendly conceivable food system.

Against 'unnatural' foods

One set of worries critics raise against processed food—especially, perhaps, plant-based meats—is that they are unnatural, fake, artificial. These are, of course, *different* arguments, but they can be hard to distinguish in practice. The core contention, however, is that plant-based meats' artificiality—the fact they imitate something, rather than being 'found' in 'nature'—is a mark against them.

Let us take 'naturalness' as a starting point.[22] The critic of plant-based meat has an uphill battle. First, she must explain what is natural about the favoured food, but this is a challenge. Let us remember that many agricultural practices are a few generations old—and agriculture itself is only a recent arrival relative to our species. Contemporary broiler chickens look, grow, and behave differently to wild junglefowl; contemporary dairy cows look, grow, and behave differently to wild aurochs. And this 'unnaturalness' is far from unique to pastoral agriculture. *Brassica oleracea* is near-unrecognizable in cultivation as kale, sprouts, kohlrabi, broccoli, cauliflower, and more. Horticulturalists bred these 'unnatural' plants from the wild cabbage—yet, thankfully, food activists do not condemn broccoli as 'unnatural'.

Second, the critic must explain what is so *un*natural about plant-based meat. This explanation will presumably vary from product-to-product. Perhaps—I make no commitment—it is relatively easy to explain what is unnatural about Impossible Foods producing haem. But it is not so easy to see why making, battering, and frying seitan (to make plant-based chicken) is unnatural—unless we are condemning *any* food preparation as unnatural.

If we are, that underlines the magnitude of the critic's third task, which is to explain what is so good about naturalness. Lots of apparently natural things are bad (disease, starvation, suffering ...), while lots of apparently unnatural things are good (literature, music, scholarship ...). It is not clear that we *should* be concerned about what is and is not a 'natural' food, diet, or food system. What matters is whether a food system can provide the variety, quantity, and quality of food needed, while respecting rights and minimizing harm. If achieving this is 'unnatural', then, frankly, to hell with nature.

This leads us to fakery. It is possible that some of the objection to plant-based meat's fake-ness is not that it is *unnatural*, but that it purports to be something that it is not. Let us be clear that we *should* object to plant-based meats insofar as they deceive consumers. Although there is no evidence that this is happening—producers of plant-based meats are pleased to make clear

---

[22] This exploration builds upon my discussion in Milburn 2022a. See also Siipi 2008; 2013.

that their products are not slaughter-based—those with a financial interest in slaughter-based meats have levelled this objection at plant-based meats. (We are lucky to have food producers *so concerned* to protect members of the public from misleading claims …)

In the zoopolis, there would be reason to oppose the production of plant-based meat deceptively presenting itself as slaughter-based meat—even beyond the general wrong of deceiving consumers, such deception raises the false-messaging (and animal-harm-normalizing) worries discussed earlier. But this does not give us a good reason to oppose plant-based meats *generally*.

Might there be reasons to be worried about the fakeness of plant-based meats even if all concerned are aware that plant-based meats are not slaughter-based meats? Perhaps there is a suspicion that there is something disrespectful in trying to imitate a food. The claim here is not the already-discussed worry about imitating parts of animals' bodies, but in imitating food that, in some sense, 'belongs' to someone else. Might there be something disrespectful, for example, in trying to imitate a favourite dish from a restaurant rather than patronizing the restaurant? Well, the restauranteur may complain. But who is the victim in the case of imitating, rather than using, slaughter-based meat? Certainly not the animals. And it is not clear why pastoral farmers, butchers, or similar would have any claim to a monopoly on meaty flavours or textures. Who is left? God? Mother Nature? Our ancestors? Even assuming we can disrespect such entities, it is unclear why trying to recreate their 'bounty' less harmfully would constitute disrespect. If anything, I would expect it to be *more* respectful.

Concluding remarks

Plant-based meats offer an easy way for meat to be compatible with animal rights—assuming we grant that plant-based meat *is* meat. Nonetheless, there are arguments that critics deploy against plant-based meat. I have demonstrated that these arguments do not provide a compelling case for the zoopolis to ban plant-based meat, although they do provide some important considerations for a zoopolis deciding in what sense it should *support* plant-based meats.

This chapter has focused on *meat*. But there are lots of animal-based foods that are not meat. As it happens, existing plant-based milks generally do not seek to closely replicate slaughter-based milks. But plant-based dairy *products* certainly do. Producers of plant-based cheeses, for example, seek to provide close replicas of camembert, feta, and the rest. And similar is true for eggs and egg products.

Many of the ethical issues raised by these products match those raised by plant-based meats. For example, they, too, may be highly processed, and they, too, may send mixed messages. But many of the arguments applying to meat will not apply to them. Tellingly, for example, although Rebekah Sinclair launches the meat-as-food objection against plant-based meats, she does not extend this to plant-based eggs and milk. These foods, Sinclair tells us, 'enact edibility differently' to plant-based meats, as they are 'referents' to eggs and milk, which 'do not imply a necessary animal death' (Sinclair 2016, 231–232).

Conversely, there may be other arguments against (plant-based) dairy and eggs. I return to these matters later in the book. Although the focus will not be on plant-based products specifically, I address ethical issues raised by dairy in Chapter 4, and by eggs in Chapter 6. A full analysis of plant-based versions of 'animal' products, then, will emerge as the book advances.

4
A defence of cellular agriculture

On 5 August 2013, journalists gathered in London, while people worldwide watched an online livestream. All were excited to see Richard McGeown, a chef, cook a beef burger and serve it to the food writer Josh Schonwald and the nutrition scientist Hanni Rützler. McGeown fried the burger, then presented it with lettuce, tomato, and a bun, although not in a traditional hamburger shape. Schonwald and Rützler, along with Mark Post (a biologist), cut and tasted slices of the meat. McGeown reported that the burger had a 'very pleasant aroma', while Schonwald noted that the 'mouthfeel' or 'bite' was that of meat (Wurgaft 2020, chapter 1).

Why would so many watch three people share a deconstructed burger? Post led the team that made this meat. Significantly, they grew it outside of an animal—at enormous cost. They did not make this burger with a piece of an animal who was once alive. But, at the same time, it was no plant-based meat burger. Post made this from animal flesh that was never part of an animal.

This 'cultivated' meat—or *in vitro* meat, cell meat, clean meat, lab-grown meat, cultured meat, and more—is not science fiction, but a tried and tested possibility. When I began writing this book, it was unavailable to the public. By the end of 2020, it was available for sale in Singapore and available (although not for sale) in Israel. This limited availability makes it far from mainstream. But its potential for a project like the current one should be clear. Cultivated meat offers the chance for meat without slaughter. It offers the chance for values associated with slaughter-based food systems—culturally, religiously, and interpersonally valuable foods; access to diverse and valued flavours; good work with food; and more—but without rights-violating slaughter industries.

Post's team took cells from living animals and grew them into strips of flesh. Over time, the team gathered enough cells to produce a mince, which could in turn be coloured, seasoned, and turned into a burger. This process is far from ethically uncontroversial. For example, Post's team utilized foetal bovine serum (FBS) as a growth medium—that is, the 'feed' for the cells. Not only is FBS a by-product of slaughter, but it is a by-product of a particularly disturbing slaughter practice. The raw ingredient is blood taken from cattle foetuses.

Post was not the first to envisage, or produce, cultivated meat.[1] Nonetheless, the London press conference marks a watershed moment in the development of cellular agriculture (that is, agriculture at the level of the cell). It proved, in the context of much public scrutiny, that producing cultivated meat (culinarily approximating slaughter-based meat) was technologically possible. Since then, the technology has rapidly developed. A wide range of start-up companies have attracted significant venture-capital investment, supported by various research and charitable organizations (Stephens, Sexton, and Driessen 2019).

Importantly, the cost of producing cultivated meat has plummeted, although technological challenges remain: for example, while producing a narrow film of cells is relatively easy, producing 'thick' tissue is trickier; while there are non-animal-based growth media on the market, finding one that is as affordable and effective as FBS is an ongoing process; and, to produce cultivated meat at scale, scientists must design and build bioreactors unlike those that have gone before. This, however, is not a chapter about the state of the cellular-agricultural industry or its technologies. As such, I will not dwell on these technological challenges except insofar as they are relevant to the ethical and political considerations explored.[2]

Before beginning, however, it is worth noting that the cellular-agriculture industry incorporates more than cultivated meat. Cellular agriculture is about growing products at the cellular, rather than organism, level. Thus, plant-based meats presently on the market incorporate the products of cellular agriculture. As noted in the previous chapter, the Impossible Burger contains haem, which is synthesized by Impossible Foods using genetically modified yeast. Other products of cellular agriculture presently available to (North American) consumers are Perfect Day's dairy products. Perfect Day produce proteins found in cows' milk, and thus dairy products without cows. Both companies use 'precision fermentation', and, although the term is novel, the technology has had important food-related uses for decades. For example,

---

[1] John Miller (2019) identifies hundreds of literary references to something like cultivated meat going back as far as 1880, with Mary Bradley Lane's science fiction novel *Mizora*. The method used to produce cultivated meat builds upon technology used for medical applications for decades. Scientists were using it to produce meat by the early 2000s, when NASA-funded experiments explored growing meat for deep-space astronauts.

[2] Readers interested in a more holistic introduction to the cellular-agriculture industry could consult, e.g., Shapiro 2018; Mellon 2020; Wurgaft 2020.

most rennet in hard cheeses is the product of precision fermentation, and not sourced from calves' stomachs (Newman 2020).

I address the ethical challenges raised by cultivated milk (and precision fermentation) later in this chapter. Importantly, the challenges sometimes differ from those faced by cultivated meat.

Other potential applications for cellular agriculture are almost limitless. Groups and individuals aiming to produce egg-based, honey-based, collagen-based, and insect-based foods/additives at the cellular level have emerged in recent years. Indeed, cellular agriculturalists have produced some surprising products: Geltor, as a proof-of-concept, reportedly produced Mastodon gelatine. These products are making their way into the market, although often not in food. This is the case with Geltor—their cell-cultured (non-Mastodon) collagen is available in cosmetics.

In this chapter and Chapter 5, I explore the ethics of cellular agriculture. In this chapter, I argue that cellular agriculture could be compatible with respect for animal rights. Thus, cellular agriculture could be present in a just state. I also argue that we do not have major ('merely') moral reasons to object to it.

In Chapter 5, I present a positive case for cultivated meat's place in the zoopolis by pointing to its potential to realize goods explored in Chapter 1. As a part of this exploration, I deal with some unfinished business from the present chapter about the place of animals in cultivated meat production. I argue that cultivated meat would be a *desirable*, not merely *permissible*, part of a just food system. But more on that later.

Cultivated meat

In this section, I rebut several arguments suggesting that cultivated meat is not compatible with animal rights. These arguments contend that although cultivated meat avoids some of the rights violations associated with slaughter-based meat, its production still involves rights violations.

Many—perhaps all—of the arguments explored in Chapter 3 apply to cultivated meat as well as plant-based meat. Endorsing cultivated meat could send mixed messages. Eating cultivated meat could problematically reinforce the idea of meat as food, or be symbolically disrespectful. Cultivated meat might be bad food, because it is meat, it is unnatural, or it is processed. These are all important challenges—but I take it that I have responded to them. Here, I address challenges more specific to cultivated meat.

Animal ingredients

Let us begin with the challenge that cultivated meat contains animal ingredients, and thus cannot be consistent with animal rights. The animal ingredients mean that—on my understanding of veganism (Dickstein and Dutkiewicz 2021)—cultivated meat is not suitable for vegans. But we should be careful not to conclude that *therefore* cultivated meat is not compatible with animal rights. There are some foods not 'suitable for vegans' that are compatible with animal rights.

That said, humans acquire almost all animal ingredients eaten today unjustly. This claim, borrowed from existing animal-rights theory, is a starting point of this book. As such, while the inclusion of animal ingredients does not rule out cultivated meat, it does give us reason to pause. Two animal ingredients in cultivated meat are particularly worthy of note: FBS and animal cells. Let us consider these in turn.

The acquisition FBS, as indicated above, certainly violates animals' rights: it involves the killing of pregnant cows, who are rights-bearers, and the exsanguination of cattle foetuses, who could be rights-bearers. At risk of understatement, it is difficult to imagine a rights-respecting route to FBS. In the zoopolis, cellular agriculturalists must cultivate meat in a non-animal growth medium. The challenge of developing this—a medium that is non-animal-based, effective, and inexpensive—is one of the major technical hurdles faced by the nascent industry. But it is one that is collectively acknowledged by the industry as a hurdle they must overcome. Suffice it to say that cultivated meat in the zoopolis would not utilize FBS, but that FBS is far from championed by real-world cellular agriculturalists.

We could make similar observations about other animal ingredients in cultivated meat. For example, cellular agriculturalists sometimes use collagen in cultivated meat production. This is a slaughter by-product, although its use in cellular agriculture is 'practically never mentioned' (Woll 2019, 775). Naturally, slaughter by-products should be opposed (as they fund rights-violating industries), and would not be possible in the zoopolis (where there would be no slaughter).

Cellular agriculturalists, however, can create collagen. Indeed, animal-free collagen produced by the already-mentioned Geltor has been commercially available since 2018. Another company, Provenance Bio, aims to produce collagen specifically for the cultivated meat industry. (Both Geltor and Provenance Bio produce collagen through precision fermentation. I consider precision fermentation below.) There is no reason to think that Geltor's Collume—or another animal-free collagen product—could not be an ethical replacement for slaughter-based collagen in cultivated meat production.

What about the use of animal cells to produce cultivated meat? This is much more fundamental to producing cultivated meat than FBS or collagen. It also sharply separates cultivated meat from plant-based meat, which does not involve animals. We saw in Chapter 3 that imitating parts of animals' bodies is not inherently rights-violating. But could *using* parts of animals' bodies (like cells!) be inherently rights-violating? Echoing earlier-cited arguments (Turner 2005, 4–6), perhaps the use of products of animal bodies in the production of cultivated meat means representing animals as mere resources, and thus disrespecting their rights.

Assuming, again, that representing animals as mere resources *would* be rights violating, we must ask if the use of parts or products of animals' bodies does inherently involve treating animals as mere resources. (Parts *or* products. We could call cells either.) This is a different argument to the one I explored in Chapter 3, which asked whether *imitating* parts or products of animal bodies might be inherently rights violating. But the same shape of argument, I think, rebuts it. We *use* parts or products of human bodies for all kinds of reasons, yet we do not take this to mean that these humans are (or that we view them as) mere resources.

Breastmilk provides an important first example. Humans share breastmilk in all kinds of circumstances, including (of course) between mother and child (cf. Gaard 2013). We donate or sell organs, blood, bone marrow, and more before and after death to lengthen or dramatically improve human lives. We use human hair to make wigs. We compost or fertilize with human urine and faeces—while perhaps the pursuit of only eccentrics in the west, human waste can be an important agricultural aid in developing countries. Those who want children trade in donated (or sold) semen and eggs. Indeed, the 'donation' and 'consumption' of semen is arguably implicit in many sex acts involving men. Scientists and medics use whole human bodies or parts of them for training and research. André Tchaikowsky's skull, bequeathed to the Royal Shakespeare Company, becomes Yoricke's in David Tennant's hands.

These uses of human bodies (or products thereof) raise ethical issues. But, surely, it is not the case that every one of these uses is rights-violating because they represent humans as mere resources. Thus, we can, in principle, permissibly use the parts or products of the bodies of rights-bearing beings without treating these beings as mere resources. The argument against cultivated meat that we are exploring thus fails.

I have certainly not shown that *any* use of rights-bearing beings' bodies (or products thereof) is unproblematic. Instead, the question of how we might permissibly source (products of) animal bodies—the case in point being cells—warrants considerable thought. This must wait until the next chapter. For now, let us allow that it is demonstrated that the fact that cultivated meat contains products or parts of rights-bearing beings' bodies does not *itself* prove the injustice of producing cultivated meat.

Historical injustice

A critic might concede that cultivated meat *could* be produced in a way that is respectful of animals, but observe that the development of cultivated meat has *already* required abuses. Regrettably, for instance, Post's team grew the first cultivated burger using cells 'from an animal already destined for the slaughterhouse' (Schaefer and Savulescu 2014, 194). It also had egg powder added (Abrell 2021, 43), and McGeown fried it in butter (Wurgaft 2020, 12). And, as noted, Post grew the burger using FBS. What should we make of this?

First, a history of abuse is not unique to cultivated meat. For example, scientists test the safety of any number of foods using animals. Consequently, if vegans are committed to rejecting *any* food we can link with rights-violating animal testing, they are going to struggle to find much to eat. Food scientists have used animal models to test the safety of potatoes, parts of potato plants (berries, seeds, etc.), and chemicals found in potatoes (such as glycoalkaloids). It would be bizarre, I suggest, to use this fact to argue that there should be no potatoes in the zoopolis. If, in the zoopolis, there is a problem with the veganic farmer feeding her community with potatoes, it is not the fact that a scientist once injected chemicals from potatoes into mice.

Of course, this is not to say we should support animal testing. The zoopolis would surely ban animal testing like this.[3] And, today, there are good (but perhaps defeasible) reasons to oppose the purchase of products routinely tested on animals, or produced by companies continuing to develop products through animal testing. But it is wrong to think that we 'taint' food with disrespect for evermore by producing it in disrespectful ways today. Equally, imagine farmers exploit migrant workers in the picking of raspberries on British farms. This should not mean that I think twice before eating raspberries grown in my own garden. Quite the opposite. Similarly, that raspberry producers *once* exploited migrant workers would not mean that raspberries were absent from the ideal food system.

---

[3] Maybe not all testing. Compare Martin 2022.

But perhaps we can draw a distinction between foods that we can (in some way) merely *link* to rights-violating practices (like potatoes), and those that food scientists *developed* using rights-violating practices (like cultivated meat). Afterall, we would have potatoes without this animal testing, and raspberries without this worker exploitation. We would not have cultivated meat without the harmful things that went on in its development.[4]

The problem with this argument is that any number of dishes and cuisines—as well as individual foodstuffs—have developed with a history of animal abuse. If I flick through the first cookbook on my shelf—2019's *Bish Bash BOSH!*, by Henry Firth and Ian Theasby—I find the Turbo Tortilla ('our take on the classic Spanish omelette'), Grilled Cheese Sandwiches (including 'our recipe for dairy-free cheese'), and—a vegan favourite—Pulled Jackfruit. Firth and Theasby are vegans. Although not as vocal about animal ethics as some others, they do not support or endorse animal agriculture. Nonetheless, their Turbo Tortilla would not be possible without abuse of chickens (so there could be eggs for omelettes), their Grilled Cheese Sandwiches would not be possible without abuse of cows (so there could be milk for cheese), and their Pulled Jackfruit would not be possible without the abuse of pigs (so there could be pigs' bodies for pulled pork).

Nonetheless, their food might seem morally innocuous. It would be surprising if we had to conclude that none of these dishes could have a place in the zoopolis. Indeed, the thought suggests a dystopian level of state surveillance and control—how, exactly, would the state prevent people from making a (vegan) Grilled Cheese Sandwich, given that the ingredients would be available? Perhaps, then, the argument is not that the zoopolis should *ban* these dishes, but that it would be *immoral* to produce them. But that, too, would be surprising. My hypothetical opponent seems to claim that blending cashews in *just such a way* (to make the 'cheese' for the Grilled Cheese Sandwich) is immoral—even while eating cashews (or blending them in a *different* way) is not. That does not seem plausible.

It is hard to imagine what food systems containing no foodstuffs that owed their development to historical animal abuse would look like. Let us not forget that (literally) from prehistory to present, humans have used animals for agriculture and foraging—think of oxen dragging ploughs. Let us not forget that (literally) from prehistory to present, humans have made animal products part of almost every cuisine. Deliberately or otherwise, we have designed dishes, diets, and the institutions of eating to complement and contain animal

---

[4] Might cultivated meat have come about via a different route? This brings us to the complicated world of modal possibilities. This is not a path we need to tread.

products. Not only do I not think we could have a food system without products owed to animal abuse, but I am not sure I could even conceive it. What would it include?

Consequently, I suggest that the appropriate way to respond to this historical abuse is to condemn it, and perhaps make amends. (What would this making of amends look like? That is an important question that I cannot answer here. Presumably, it would involve *more* than simply extending robust rights to animals. But that would be the first step.[5]) But refusing to utilize potentially valuable foods—grains, a Turbo Tortilla, cultivated meat—because of historical wrongdoing does not help the animals to whom we should (perhaps) make amends, even while it is damaging to those who value these foods.

In response, the critic of cultivated meat may note that while no one will benefit from our not utilizing these foods, we nonetheless do wrong in benefiting from an injustice, as anyone who enjoys cultivated meat does.[6] I deny that there is such a wrong. Imagine a terror attack forces you to take cover in someone's house. The two of you would not otherwise have met, but you remain in contact after the attack, and—speaking euphemistically—one thing leads to another. You end up living long and happy lives together.[7] Clearly, these long and happy lives are good—for you, them, people around you. But you owe them to a terror attack, which (we can stipulate) involved injustice. It would, I contend, be absurd to suggest that you do something all-things-considered (or even prima facie) *wrong* in living a long and happy life with a person you love because in so doing you benefit from an injustice. Thus, I suggest, there is no *inherent* wrong in benefiting from an injustice.

I conclude that although there have been abuses in the development of cultivated meat, this is not a good reason for us to not produce it. For the sake of clarity, it is worth stressing that this does *not* entail that we should tolerate current injustice in the pursuit of cultivated meat.[8] The fledgling cultivated meat industry could be more respectful of animals, and we can consistently critique the current industry while recognizing that cultivated meat could be a part of a just food system *in the future.*

---

[5] For work beginning the conversation of what duties we owe to animals given this history of abuse, see Mosquera 2016 and Scotton 2017.

[6] Does this apply when it is the victims of the injustice benefiting from the existence of the putatively unjust foodstuff? What about when it is the descendants of the victim of the injustice? (And when we are talking about descendants, do we encounter non-identity problems?) These questions are difficult.

[7] I borrow this from a similar thought experiment used by Tyler Doggett (2018), who borrows it from Garrett Cullity via Christian Barry and David Weins (2016).

[8] Although it is at least plausible that we should. Compare Abbate 2020.

The false-hierarchy objection

In the eyes of its critics among animal ethicists, the prospect of the development of cultivated meat creates (or sustains) an 'us' and a 'them' (Donaldson and Kymlicka 2011, 152). *Us* humans (plus companion animals, charismatic megafauna …) and *those* historically farmed animals: An inedible us, an edible them. This two-tiered system is anathema in the zoopolis, where humans and animals are equal co-members. We can call this the *false-hierarchy objection* to cultivated meat: A food system incorporating cultivated meat reifies, reinforces, or instantiates a false moral hierarchy.

There is an easy way to break down this binary. As critics typically present the argument, the false-hierarchy objection pushes us towards the idea that we break down the difference between the inedible *us* and the edible *them* by affirming that both *us* and *them* are *in*edible.

However, we could as easily break down the binary by affirming that both *us* and *them* are *edible*. This might sound like a 'levelling down'. It brings the favoured class 'down' to the level of the disfavoured, rather than bringing the disfavoured 'up' to the level of the favoured. But, as we have seen, recognizing someone as the potential source of a consumable product is not necessarily disrespectful. We do not think of blood donors as second-class citizens. Indeed, strikingly, recognizing someone as the source of a consumable product can mean the opposite. We are more comfortable with organs and blood from humans than from animals (in part) because we think of humans as 'higher'.

My proposal is this. If we are creating meat without harm, why stop at animals conventionally eaten in the west? Why not include animals eaten in other (cultural, geographic, temporal) environments? The dogs, cats, and horses beloved by westerners find a place in non-western (and some western) diets. Charismatic—intelligent—megafauna are or have been important parts of certain cuisines. Dolphins are part of Japanese and Faroese cuisine. Humans have eaten elephants since the early days of our species. Central African hunters target gorillas and chimpanzees for bushmeat. Cellular agriculturalists could produce meat from these animals for those who value access to it. Better, cellular agriculturalists *should* produce meat from these animals for those who value access to it.

Of course, the big dividing line is not between beloved animals and less-well-loved animals. It is between humans and animals. We *could* create cultivated meat with from human cells. Should the zoopolis permit cannibalism? In short, yes. Why not?

Cannibalism emerges in a surprisingly large number of discussions of cultivated meat (Wurgaft 2020, chapter 16). Sometimes, critics present it to evoke a negative association; if we associate cultivated meat with cannibalism, that is a mark against it. Of course, this is not much of an argument. I contend that there are reasons to welcome the prospect of cultivated human flesh, roughly analogous to many of the reasons we have to favour cultivated animal flesh.

First, and if nothing else, plenty of people are sufficiently *curious* about cannibalism that they would be keen to try (cultivated) human flesh. The popular website Vice reports that, in 2016, an American man got the opportunity to legally serve human flesh after doctors amputated his foot: 'I invited 11 people. I said something like, "Remember how we always talked about how, if we ever had the chance to ethically eat human meat, would you do it? Well, I'm calling you on that. We doing this or what?" Ten said yes' (Mufson 2018). (Interested readers can find the recipe online. According to the amputee, his flesh 'had a very pronounced, beefy flavor' (Mufson 2018).)

The amputee said that the meal was a valuable 'bonding experience', and that it helped him achieve closure on the accident that led to the amputation (Mufson 2018). Now, this does not show that ethical cannibalism will always involve bonding experiences or personal closure. But it does show that valuable personal and interpersonal experiences can be at stake in the eating of human flesh.

I do not claim that the anonymous amputee's ten out of eleven rate of interest will be representative of a wider rate of interest, but the fact that placentophagy—placenta-eating—is an enduring, cross-cultural, and (relatively) common practice suggests that there is at least some desire to engage in cannibalism.[9] (More on placentophagy in Chapter 6.) Is it hard to imagine that foodies, thrill-seekers, or just those who like the 'vibe' might frequent a restaurant called The Ultimate Taboo, relishing its surprisingly meaty menu? If so, why should the state not leave these people to enjoy their taboo-busting evening, just as it rightly leaves swingers and BDSM enthusiasts to enjoy theirs?

Second, perhaps(?) more seriously, certain traditional cultural/religious practices and cuisines use human flesh. There are few cannibalistic cultures remaining in the twenty-first century.[10] Historically, many cultures

---

[9] Indeed, the desire is so strong for some that they seek out 'consenting' individuals to kill and eat. Armin Meiwes, a German who killed and ate Bernd Brandes in 2001, is currently serving life in prison.

[10] The Aghori, an Indian Saivite sect, present a complicated example; the Wari', in the Amazon, regularly engaged in cannibalism in living memory; the Korowai, of New Guinea, *may* continue to practise cannibalism.

have regularly engaged in cannibalism, including some indigenous peoples of the Americas.[11] Whether or not a given cultural, religious, or ethnic group engages in cannibalism today, it is plausible that individuals may find deep cultural significance in cannibalism, just as many individuals find cultural significance in (conventional) meat-eating.

Compare: Whale meat is deeply significant to the Makah, even though they have not regularly engaged in whale hunts for around a century (Kim 2015, chapter 7). While it is right that arguments about the cultural significance of cannibalism cannot override concerns about human rights, we should be more cautious when it comes to *harm-free* cultural practices, even if those practices seem unsavoury. It is my contention that the consumption of cultivated human flesh could be such a harm-free cultural practice.

Now, I am not claiming that the entirety of the significance of cannibalism for (say) Aztecs is solely about a particular taste, just as the entirety of the significance of whale meat for the Makah cannot be. (For the Aztecs, human sacrifice was presumably important; for the Makah, whaling is important.) Indeed, it is neither my aim nor my place to say what is and is not important for members of particular cultural groups. All I am saying is that the consumption of human flesh could be important. Even if we cannot engage in literal human sacrifice, harm-free human flesh could prove a crucial part of the puzzle in developing or constructing a rights-respecting version of a culturally important practice.

Third, it could be that *human* flesh is particularly valuable, paradoxically, for those concerned with animals. On some occasions, I have convinced people sceptical of cultivated meat of its value for the feeding of carnivorous companion animals (e.g., Deckers 2016, 98–99). At other times, I have encountered resistance to cultivated meat even in this case. The worry is that it denigrates animals (discussed throughout this chapter), and that we should not take cells from animals (discussed in the next). The idea that we could feed animals cultivated flesh from *humans* has been, on these occasions, more warmly received (cf. Bovenkerk, Meijer, and Nijland 2020), as it can overcome these problems (it involves no denigration of animals, and no one has their cells taken without consent).

In some ways, it is not surprising that animal advocates would be open to human flesh. Indeed, a fixation on the edibility of humans is present in some 'posthumanist' literature (e.g. Plumwood 2000), indicating the significance some activists place in affirming our own 'meatiness'. Cultivated

---

[11] We might dispute many attributions, but the Aztecs, for instance, engaged in cannibalism at least *occasionally*.

human flesh provides such activists with (literal) material for their cause. At least, it provides food for thought—a reflection on why we are opposed to even (ostensibly) harm-free cannibalism (if we are) gives us reasons that we may be concerned about cultivated meat more broadly (Alvaro 2019). For instance, we saw in Chapter 3 how reflecting on the creation of plant-based human skin can help us get to the bottom of the ethics of eating harm-free meat—plant-based or cultivated.

Fourth, looking beyond meat reveals a host of reasons to explore growing human matter at the cellular level. (Indeed, cellular agriculturalists borrow technology from medical engineering—there are medical and research reasons to grow human cells.) To nod towards a later discussion, breastmilk provides an illustrative example. Many mothers, including those who cannot produce enough milk to feed infants, those concerned about breastfeeding interfering with their careers, or those who cannot safely breastfeed because of drink or drugs, could benefit from access to breastmilk developed through cellular agriculture. Various systems—from the formal to the informal—are already in place in some contexts for mothers to access breastmilk from others, but these can be unreliable or costly. For example, for handling *donated* breast milk, hospitals will pay '$50 per liter in Norway, $96–160 a quart in the United States' (Gaard 2013, 600). If hospitals had a reliable, cheap, and safe source of human milk, they could revolutionize infant medical care.

Thus, many humans (and cows) could benefit from cultivated human milk before we have even entered the world of gastronomical experimentation. When I first made this argument (Milburn 2018), I predicted and hoped that groups bioengineering breastmilk would be forthcoming. Since that time, BIOMILQ, TurtleTree Labs, Harmony, Helaina, and other start-ups exploring exactly that possibility have emerged. The development of cellular agricultural technologies (and their inclusion in just food systems) is as much about benefiting humans as benefiting animals.

In short, we can undermine the false-hierarchy objection by simply embracing cannibalism—or, specifically, the production and consumption of cultivated human flesh. In fact, we have good reason to do this anyway. Might there be other compelling arguments against cannibalism? Perhaps. But, I conjecture, as we explore the reasons we might be worried about cultivated animal meat—including those addressed in the previous chapter—we will come to terms with why we should not be too worried about the prospect of cultivated human meat, either.[12]

---

[12] Are there reasons to be worried about cultivated human flesh that do not apply to cultivated (say) pig flesh? Lots of these arguments reduce to human supremacism, but there are at least three exceptions.

Cultivated milk

I return to cultivated meat in Chapter 5. For now, let us turn to consider cellular agriculture beyond meat. In an important sense, it is non-meat cellular agriculture that has established itself first. As already noted, for example, Impossible Foods's haem and Geltor's collagen have been available for years, and fermentation-based rennet has been available for decades. But more striking than these cell-cultured additives is the existence of cultivated milk. Perfect Day is an American company that has been producing cultivated dairy proteins since 2019—available to consumers in ice cream, milk drinks, and more. Cultivated dairy products are thus available (if likely unfamiliar) to Americans. Other companies focus on other dairy products. The German company Formo, for example—although their products are not, at time of writing, commercially available—produces cheese.

The technologies of precision fermentation used by makers of cultivated milk are simpler than the technologies used for cultivated meat. Cultivated dairy producers genetically modify non-animal cells (say, yeasts) to produce organic compounds (say, casein, a milk protein) via fermentation. The producers thus do not need animals. At most, some understanding of animal genetic makeup is required. (Unsurprisingly given their commercial significance, cattle genetics are already very well-known to humans.)

At the risk of oversimplifying, the process uses brewing technologies that have been relatively familiar to humans for millennia; the only high-tech part is the initial genetic modification. The Good Food Institute carefully distinguishes *traditional fermentation* from *biomass fermentation* and *precision fermentation* technologies. An example of the first is turning soybeans into tempeh. The second creates foods' bulk—food scientists ferment fungi, for example, to produce Quorn's plant-based meats. It is the third with which I am presently concerned: genetic modification allows for the creation of cell-level factories to produce particular proteins (or similar). But even the genetic technologies are accessible compared with the technologies used for cultivated meat. Indeed, some of the first people to explore cultivated dairy

---

First, cannibalism is disgusting, even compared to pig-eating. I respond to this at length in Milburn 2016. Second, cannibalism is unhealthy, even compared to pig-eating. Even if unhealthfulness is a good reason to ban a food (compare the discussion in Chapter 3), this does not serve as an argument against producing human flesh. We could still use such flesh in, for example, pet food. Third, the eating of human flesh in this way sets us on a slippery slope towards the eating of human flesh in other ways. Philosophers normally consider 'slippery slopes' bad arguments, for straightforward reasons—we can stop things from sliding down slopes with appropriate barriers. In this case, the 'appropriate barriers' are human rights. As humans have rights, we can only go so far 'down' the 'slope'.

were hobbyist biohackers contributing to the Real Vegan Cheese project (see Wilbanks 2017).

Despite the relative availability of cultivated milk compared to cultivated meat, ethical conversations about cellular agriculture have focused on the latter. And many of the critiques of cultivated meat (including those already explored) could extend to cultivated dairy. For example, critics might contend that cultivated milk is symbolically disrespectful, unnatural, or highly processed, or argue that endorsing cultivated milk sends mixed messages. As explained in Chapter 3, these objections fail when it comes to plant-based meat; similarly, they fail when it comes to cultivated milk.

Additionally, the false-hierarchy objection to cultivated meat explored in the previous section could be applied to cultivated milk—but, as I argued, the possibility of cultivated breastmilk is actually a compelling argument *against* the false-hierarchy objection, and multiple companies aiming to produce cultivated breastmilk have emerged (albeit not always via precision fermentation).

In this section, I address further arguments that are specific to cultivated milk, but which are less relevant to the production of plant-based or cultivated meat. First, I address challenges focusing on the wrong of consuming *milk*—considering, too, to what extent they could serve as objections to including milk within a food system.

Second, I turn to consider objections to *cultivated* milk. Next to the methods used to cultivate meat, it may be tricky to imagine that precision fermentation is objectionable. But I note two worries. The first is that precision fermentation, unlike cultivated meat, involves genetic engineering. The second is that precision fermentation, unlike cultivated meat, has no room for animals.

Milk is not food

In Chapter 3, we explored the challenge that plant-based (and, by extension, cultivated) meats problematically reinforce the idea that meat is food. Critics could offer a parallel challenge to milk. Cultivated milk, on at least some understandings,[13] *is* milk. If we should not eat dairy products, then we should not eat cultivated dairy. Interestingly, though, some of the reasons we might

---

[13] If we define milk relative to culinary purposes or biochemical makeup, cultivated milk (in principle, at least) is milk. If our metaphysics of milk depends upon milk having a particular causal relationship with an animal's body, cultivated milk (probably) is not milk. Compare the discussion of the metaphysics of meat in Chapter 3.

object to conceptualizing milk as food differ from the objections we might have to meat as food. They are thus worth reviewing.[14]

First, take metaphysics. If milk is not food, perhaps it has no part in a food system. But I contend that there is something confused about redefining milk, *a priori*, as non-food, whatever account of the metaphysics of food we endorse. Milk exists solely as food; it is thus different from meat, which, if slaughter-based, exists first as the body of an animal. Further, denying that milk is food suggests that infants drinking their mothers' milk are consuming something that is not food. This seems straightforwardly incorrect.

Second, activists endorse slogans to the effect that cows are 'not your mother', meaning that what they produce is 'not your milk'. This points to a moral claim about the wrongness of consuming milk (that is not from your mother). Alternatively, activists will frame milk as food for babies (human or otherwise); thus, milk drinkers are unethically stealing from babies. We can imagine a political variation of these slogans, claiming it is unjust to take milk, as it is someone else's food, meaning milk (exempting the breastmilk that mothers provide for babies) has no place in the zoopolis. (Such a charge is, of course, independent of the correct charge that the slaughter-based dairy industry is unjust because of the rights of cows and calves.)

The sentiment behind these slogans is admirable, but, as arguments for ethical principles, they go wrong. Greta Gaard (2013) explores examples of milk sharing beyond mothers sharing milk with their own infants. For example, in Latin America and Africa, women share nursing requirements to help each other and children, while in the west, mothers create social-media-based spaces to arrange the sharing of milk with the children of others (Gaard 2013, 601). And sharing goes beyond human-mother-to-human-infant sharing; Gaard points to the human suckling of 'pigs, dogs, monkeys, and bear cubs in precolonial Polynesia, the forests of South America, and the hunter-gatherer societies of Southeast Asia, Australia, and Tasmania' (2013, 599).[15]

What is the significance of this sharing? We have already seen how claims about the inherent wrongness of utilizing resources from another's body impact a range of apparently innocuous practices. But I contend that these examples show, further, that claims specifically about the inherent wrongness of consuming *milk* (including milk that is not from one's own mother) impact a range of apparently innocuous practices.

---

[14] The remainder of this subsection borrows from Milburn 2018, 271–274.
[15] Mothers sometimes sell breastmilk as food for adults (as milk, cheese, ice cream, etc.) although, Gaard reports, it fares poorly (2013, 602).

What is more, the autonomous sharing of breastmilk holds a place in human culture and history. In classical mythology, a wolf suckled Romulus and Remus, while in Catholic hagiography, the Virgin Mary breastfed St Bernard of Clairvaux in a vision. John Steinbeck's *Grapes of Wrath* closes with Rose of Sharon offering her breast to a starving man. Traditional and alternative medicines have made, and still make, use of human breastmilk.

Stories about the 'sharing' of one's own flesh are harder to take seriously outside of life-and-death emergencies, extreme (sexual or otherwise) fetishes, or highly idiosyncratic religio-spiritual rituals (Wisnewski 2014). Thus, an ethical rejection of milk-as-food seems, unlike a rejection of meat-as-food, to impact upon a wide range of autonomously undertaken, intuitively innocuous practices. For this reason, I consider it suspect.

Third, vegans sometimes present milk as disgusting. In Chapter 3's exploration of disgust and meat, I argued that disgust does not provide a good reason to ban foods. But, further, I believe we should reject the contention that milk is disgusting. Yes, there is something disgusting about anything produced through the means utilized by the slaughter-based dairy industry, just as there is something disgusting about products created through practices exploitative of humans. If one is disgusted by milk *itself*, however, there are troubling consequences for one's view of mothers—human or otherwise—offering their own milk to their children. This emphatically is not disgusting, but it is hard to see how we could reconcile this claim with the idea that *milk* is disgusting. (One must also ask whether we can separate disgust directed at milk from implicit misogynistic aversion to women's bodies.[16])

Fourth, vegans identify claims about the healthfulness of milk for individuals as reasons to reject the idea that milk is food. My response to this echoes my response to the claim that plant-based meats are unhealthy. First, it would be an uphill battle to claim that milk is *so* unhealthy that it is something other than food. Second, proposing that milk consumption is unethical or should be illegal because it is damaging to health seems to be the worst kind of paternalism. (Indeed, opponents of milk consumption on health grounds should welcome cultivated milk, as it reduces some of the health risks presented by slaughter-based dairy *production*, including those relating to pollution.) Third, cultivated milk offers the potential for fine-tuning, producing dairy products that can minimize particular health risks.

---

[16] I thank Sue Donaldson for this observation.

Fifth, milk consumption is a racialized phenomenon. As Gaard explains,

> Populations that have a historical practice of milking domestic animals ... have retained the enzyme (lactase) that digests lactose sugar in milk, far beyond childhood; however, the majority of the world's populations lose the lactase enzyme by the age of four, and thus lactose intolerance is common among Vietnamese, Thai, Japanese, Arabs, Israeli Jews, and African Americans, Native Americans, Asian Americans, and Hispanic Americans.
>
> (Gaard 2013, 608)

Consequently, activists frame the foregrounding of milk in the dietary advice of western, predominantly white, nations as displaying a problematic ethnocentrism. Indeed, the far right explicitly champion dairy, with racists using statistics relating to lactose-intolerance as 'evidence' of white supremacy, and references to milk as coded racist messages (Stănescu 2018). Maneesha Deckha even calls for a centring of anti-dairy in vegan activism:

> First, a focus on dairy better reveals how veganism can be read as a counter-force to settler colonialism in North America ... demonstrated through a respect for animals and their families (particularly for mother-child bonds) and by highlighting the constitutive role of cows and dairying in colonial expansion in North America, as well as by underlining the otherwise Eurocentric nature of dairy milk consumption. Second, the focus on protesting the dairy industry—unencumbered by the question of Indigenous rights to hunting, fishing, and trapping—can help us appreciate how both veganism and many Indigenous societies in Canada disavow human exceptionalism and the sharp species divide between humans and animals that subtends it. This divide was, moreover, a foundational settler colonial ideological narrative and is central to the staggering objectification and commodification of animals today.
>
> (Deckha 2020, 249–250)

It is undeniable that state agencies must be aware of racist and colonial framings and uses of nutrition science when drafting food policies. But I contend that these concerns about milk should not make us reluctant to endorse the place of cultivated milk in the food system of the zoopolis.

Why? Dairy consumption is not *inherently* racist or colonial. If it were, a great many people—including committed anti-racists, anti-colonialists, and both racialized and indigenous people—would be engaging in racist or colonial practices. This is a bullet we could bite. But, if this is a bullet we *should* bite, we can return to our earlier considerations about the meanings of meat (see Chapter 3). *If* dairy consumption is inherently racist or colonial, better—given potential positive associations of dairy, and the importance of dairy to many—to change the meanings of dairy, so that its consumption is no longer racist or colonial.

This leads to my second response. Cultivated milk offers a powerful potential corrective to racist or colonial messages in milk because it can undercut the racist or colonial elements of dairy. This is because cellular agriculturalists can produce it *without* the (so-called) allergens that most adults find indigestible—and thus there is no need for its consumption to be Eurocentric—and because its production does not entail the disrespect of animals or the land that the production of slaughter-based dairy involves.

Deckha's nod towards slaughter-based dairy's destruction of family bonds speaks to a sixth objection to dairy: Specifically, that it entails the objectification of motherhood and female reproductive systems—dairy is *feminized protein* (Adams 2017). But as its production does not involve animals, cultivated dairy significantly weakens this challenge. No *actual* animals have their reproductive processes or motherhood objectified in the production of cultivated milk. (Perhaps this response is too quick. I revisit feminized protein in Chapter 6.)

For now, let us conclude that these assorted worries about framing milk as food—borrowed from vegan activism and scholarship—do not provide compelling reasons for the zoopolis to ban or oppose the production of cultivated milk.

Genetic engineering

A key difference between cultivated meat and the products of precision fermentation—including cultivated milk—is that the latter involves genetic engineering. (It is not always easy to draw a clear line between cultivated meat and precision fermentation, insofar as products could draw upon both technologies, and because cellular agriculturalists also explore fermentation as a source of meat. Let us put these complications aside.) Genetically engineered food is controversial. It is perhaps more controversial in the public domain than in academic circles, although academics do raise legitimate concerns. Most pertinently, worries about the safety or outcomes of *specific* instances, products, or aims of genetic engineering may raise eyebrows.

Scepticism about genetic engineering *generally* is more associated with public responses to food technology. Sometimes, objections rest on questionable or outright fallacious grounds, such as the 'unnaturalness' of genetic engineered foods. We saw in Chapter 3 that the putative 'unnaturalness' of plant-based meat should not worry us. Might genetic engineering involve a different kind of unnaturalness?

It is sometimes suggested that the unnaturalness of genetic engineering crosses a line that domestication, selective breeding, or hybridization do not—'unnatural' though such agricultural techniques are. The difficulty (in addition to the obvious challenge of explaining the bad of unnaturalness) is in identifying what this line is, and how genetic engineering crosses it. This difficulty leaves critics offering this objection caught between 'limited and even Luddite' pro-'natural' arguments and arguments for genetic engineering's exceptionalism that are, as Rachel Ankeny and Heather Bray diplomatically note, 'difficult to assess' (2018, 103–104).

I cannot here explore the ethics of genetic engineering at length. Indeed, I take no stand on how (if?) the zoopolis should regulate genetic engineering, beyond arguing that precision fermentation's reliance on genetic engineering should be no barrier to its presence in the food system. To do this, I will turn to what is—after concerns about 'naturalness', 'playing God', and similar—the key challenge to using genetic engineering in the pursuit of food, which is that '[genetically modified] organisms that are released or used have unacceptable and unknowable risks to human and/or animal health' (Ankeny and Bray 2018, 104)—or, the health of non-animals, or the 'health' of the wider environment/ecosystem. These are (in principle) good reasons for the state to regulate or restrict genetic engineering in food production. But asking whether genetic engineering does pose significant[17] risks requires close attention to *particular* instances, uses, or deployments of genetic engineering.

To that end, I offer four sets of reasons that precision fermentation's genetic engineering is *less* worrying than some other forms of genetic engineering used in food production. Even if—on which I take no stand—the zoopolis should restrict those other forms of genetic engineering, it is not obvious that it should restrict precision fermentation.

First, as I noted earlier, precision fermentation involves the creation of cellular factories—but the factories are not, themselves, the food. Cultivated milk *itself* does not contain genetically engineered organisms. Instead, it contains proteins (or lipids, or pigments, or what-have-you) that physically replicate[18] those in slaughter-based milk. Someone committed to not eating genetically engineered organisms could, consistently, eat cheddar cheese

---

[17] *Significance* here understood as some function of the likelihood of a negative outcome and the harm that said outcome would cause. This is not straightforward.
[18] Or do not—cellular agriculturalists can also use precision fermentation to create *new* proteins, lipids, and whatnot. But they would not be genetically modified, either.

made with cultivated dairy proteins.[19] That cultivated milk is not *itself* genetically engineered has safety benefits—or perceived safety benefits. In debates about genetic engineering, we are (perhaps understandably) particularly concerned about 'what goes into our mouths', rather than in the existence of genetically engineered organisms as such (Comstock 2012, 117). Crucially, risks associated with the *consumption* of genetically engineered organisms do not apply to cultivated milk, even if they might to other foods discussed in debates about genetic modification—from rice to tomatoes to slaughter-based beef.

Second, precision fermentation's genetic engineering is genetic *editing*, which is a targeted, precise form of genetic engineering. This precision comes with advantages. It means that unexpected side-effects are much less likely. Any feared cow–yeast hybrid is not a real possibility. Genetic editing does not involve chopping out (small) parts of cows and slotting them into yeast, or injecting yeast into cows—it does not require cows at all. Indeed, at least in principle, it may not require adding anything to the genome of microorganisms that does not (or could not) appear there already. When cellular agriculturalists use a 'host' organism to create a protein,

> the instruction manual for synthesizing the protein is encoded in the host organism's DNA, either as a naturally occurring gene or as a gene introduced through engineering. Depending on the target, both engineered and non-engineered approaches may be possible. … For example, the soy leghemoglobin protein produced by Impossible Foods is engineered into a yeast host strain for efficient, scalable production. On the other hand, microalgae company Triton Algae Innovations is commercializing heme proteins that are native to their algal strains, so no engineering is involved.
>
> (Specht n.d.)

Creating non-proteins is a little more complicated, but—equally—carries no risk of creating bovine yeasts or fungal cows. While there is room for worries about the risks imposed by less precise forms of genetic engineering, genetic editing can sidestep many of them.

---

[19] If they ate supermarket-bought *slaughter-based* cheddar cheese, there is a good chance they would be eating food similarly one-step removed from a genetically engineered organism. Namely, an organism that has produced rennet. Indeed, *American* slaughter-based cheddar cheese likely contains *milk* only a few steps removed from genetic engineering. American dairy cows are frequently injected with recombinant bovine growth hormone (rBGH), an artificial version of a hormone that naturally occurs in dairy cows in small quantities. Humans produce rBGH for dairy farming using genetically modified microorganisms. *This* use of genetically engineered organisms in food production raises ethical issues insofar as we now must consider the impact on *cows*.

Consumers appreciate this distinction. The following is from a report on preliminary results of focus groups with potential early adopters of cultivated milk:

> Participants became very curious about the technology behind animal-free dairy, [Garrett] Broad said, such as whether the products were safe. One of the questions they discussed was whether dairy created through precision fermentation was like GMO food. ... Consumers were less concerned as differences between GMO food and precision fermentation were explained, and this application of fermentation was readily accepted, said Oscar Zollman Thomas, a business analyst at Formo. For precision fermentation dairy, DNA is not modified, just copied.
> (Poinski 2022)

So even those concerned with the impacts of genetic technologies in our food system could be, and are, supportive of precision fermentation.

Third, precision fermentation uses highly contained genetically engineered organisms. Precision fermentation does not require fields of genetically engineered organisms separated from 'nature' by a wooden fence. There would be no reason for these organisms to leave the brewery. There is thus no reason to believe that there would be any significant chance of them 'spreading', as there might be with—say—a field of genetically modified wheat, which might naturalize, or 'infect' neighbouring wheatfields, or otherwise impact ecosystems. The genetic modification required to produce cultivated milk will not have adverse impacts upon the 'natural' world.

At the same time, precision fermentation could have a *positive* impact on the natural world, insofar as the environmental impact of cultivated milk (through carbon emissions, water use, and so on) will be lower than that of slaughter-based dairy. (It could be lower than the production of some or all plant-based milks, too, but I make no commitment on this.) Thus, consumers generally concerned about the impact of genetic engineering on ecosystems could well support the use of precision fermentation.

Fourth, precision fermentation's genetically engineered organisms are not *themselves* rights bearers—we do not need to be worried about harming them. This is unlike some cases of genetically modified farmed animals. Nor need we associate these organisms' use with harm to non-modified animals (or ecosystems). This is unlike cases including genetically modified crops that have increased pesticide resistance, allowing the increased use of (harmful) pesticides. And, of course, cultivated milk can *reduce* harms to animals insofar as it replaces slaughter-based dairy (or even, hypothetically, animal-harming plant-based milks). Consequently, someone concerned about genetic engineering's impacts on animals could support the use of precision fermentation.

None of what I have said implies that the zoopolis would not provide oversight of genetically engineered foods. Again, I take no stand on that. But I do not think that cultivated milk's use of genetic engineering is a good reason to oppose its presence in the zoopolis, or believe that it is something that should be merely tolerated rather than supported. On the contrary, given the assorted values that people place in access to milk, the health benefits that milk can have, and the prospect of a particularly non-harmful form of milk production, there is every reason to think that the zoopolis would and should support the production of cultivated milk.

Animals—or their absence

Clean milk production requires the 'use' of animals very indirectly. Namely, it requires knowledge and data gained from the study of animal genetics and food science. The scientists with whom this knowledge originates probably did not treat animals as their rights demanded. But we have already seen that historical abuse of animals in pursuit of food should not stop us from benefiting from that food today. And, more, scientists did not study these things *so that* cellular agriculturalists could make cultivated milk. Cellular agriculturalists make use of knowledge acquired for *other* reasons.

Beyond this, however, cultivated milk does not need animals. Were aliens to rescue every cow on earth tomorrow, Perfect Day could continue to produce dairy. A food system that replaced slaughter-based milk with cultivated milk would need no cows—or dairy farmers, or dairy farms. On the face of it, this is a good thing. It is a food system without the suffering and death associated with slaughter-based dairy; none of the environmentally destructive carbon emissions associated with slaughter-based dairy; none of the extensive land-clearing and land-use associated with slaughter-based dairy.[20]

But there are negatives to losing dairy farms and farmers, echoing those explored in Chapter 1. People lose good work, and practices that are important to people cease. There is an impact on our image of the countryside—indeed, the literal *shape* of the countryside changes. Small-scale pastoral farming practices that could be environmentally beneficial are gone. And we figuratively tell some animals that they have no place in the zoopolis—animal liberation means the end of animals who *could* live good lives.

---

[20] And, as I hinted above, it is at least plausible that cultivated milk would be less harmful than some plant-based milks—although precise calculations are difficult to make.

I address associated arguments concerning cultivated meat in Chapter 5. But, for now, let us ask whether there is reason for regret if cultivated (and plant-based) milk completely replaces milk from animals.

There are at least two sets of responses to this. Both have merit. The first is that, yes, there are losses in the food system of the zoopolis relative to the current food system. But the current food system's advantages depend upon the injustice of slaughter-based dairy farming. Unjust advantages are an inevitable casualty of social progress. *Yes*, we lose a particular (say) livelihood, but we had no right to it in the first place.

Equally, child labour laws mean employers lose a source of cheap labour, and are thus minimally regrettable. But that is hardly a compelling argument against child labour laws. If justice demands that we stop treating animals (or children, or ...) *like that*, then we must stop—even if that means we must give up something valuable, advantageous, or similar. Of course, we can combine this response with the suggestion that there are alternative routes to these values. Maybe, for instance, the zoopolis could preserve good work with animals, *and* find a place for animals, *and* preserve a favoured image of the countryside, *and* champion environmentally beneficial cow/human co-living by providing support for cattle sanctuaries.

This leads to my second response. I do *not* believe that the zoopolis should ban all forms of dairy farming. On the contrary, because of the values associated with pastoral farming (indicated above, and explored in Chapter 5), I believe the zoopolis could, in principle, *champion* some forms of small-scale, *genuinely* humane dairy farming. In Chapter 5, I introduce rights-respecting animal agriculture in the context of cultivated meat. In Chapter 6, I explore the possibility of applying this thinking to egg farming, arguing that there would still be a place for egg farming in the zoopolis.

I believe we could envision a similar system for dairy farming—whether with cows, goats, sheep, or others. I take my discussion of egg farming in Chapter 6 as indicative of the farming I imagine, so I will not, here, sketch details of a rights-respecting dairy farm. Equally, I do not attempt to sketch the details of a rights-respecting duck egg farm; rights-respecting shepherding for lanolin or non-food wool products; of (returning to a thought from Chapter 2) rights-respecting apiculture for honey and other bee-derived ingredients. Each possibility brings its own puzzles; it is my hope that future work will identify rights-respecting routes to a range of animal products.

Incidentally, there are already farms beginning to approximate a rights-respecting dairy farm. 'Ahimsa' dairy farming, for example, kills no cattle—cows or calves. And while killing is not the only problem with slaughter-based dairy, it is a key concern. We can, then, put aside questions about cultivated milk based on how it will remove some of what is valuable in dairy farming. Not only is the objection by-the-by—if slaughter-based dairy farming is unjust, we have no right to those valuable things—but there is no *in principle* reason that a form of small-scale dairy farming could not, would not, and should not remain in the zoopolis.

Concluding remarks

This chapter, building upon my exploration of plant-based meats, has canvassed reasons to believe that cellular agriculture could not be a part of a just food system, finding them wanting. Importantly, it has addressed both cultivated meat *and* precision fermentation.

However, there is some unfinished business. My discussion of precision fermentation has focused on cultivated milk, when there are a range of other technologies. Some cellular agriculturalists use fermentation to produce meat, compounds found in meat (e.g., haem, proteins), or non-meat products of animal bodies (e.g. fats, gelatine). I take it that these do not raise any ethical issues that I have not covered.

Other cellular agriculturalists, however, produce animal foods that are neither meaty nor milky. Most significantly, Clara Foods, Fumi Ingredients, and The Protein Brewery are all working, in various ways, on the production of cultivated eggs. MeliBio, meanwhile, is aiming to produce cultivated honey. *These* products raise further questions. I offer further discussion of eggs in Chapter 6; although the focus will not be on cellular agriculture, it is worth bearing cellular agriculture in mind when considering the arguments. Indeed, to pre-empt, it is my view that *most* eggs in the zoopolis will be plant-based or cultivated, even if there may be some just ways to source eggs from hens.

There is also unfinished business concerning cultivated meat. This chapter has left aside how we acquire animal cells to start cultivated meat's production. Without addressing this question head-on, it is unclear how, if at all, animals fit into the envisioned cultivated meat industry. While Chapter 5 does deal with this unfinished business, its focus is on offering a *positive* case for cultivated meat. A food system containing cultivated meat, I argue, has key advantages over a meatless system. These are not merely advantages for those who like eating meat—but, and perhaps surprisingly, advantages for animals, too.

5
A positive case for cultivated meat

Critics paint cultivated meat as dystopic. They evoke a world which separates humans from animals, nature, and the food system (Fairlie 2010, chapter 15); where farmers keep animals in hellish conditions for cells (Lance 2020); where humans live off uniform, 'soulless' food (Evans 2019, 215); where amoral elites control food systems (Miller 2012, 55).

In this chapter, I offer a counter-vision, sketching how cellular agriculture—particularly cultivated meat—could contribute to a just food system. More than contributing, it could be important in overcoming many of the problems with a plant-based food system outlined in Chapter 1 *beyond* those narrowly concerned with access to animal products. The envisioned food system could secure access to food, provide good work, and sustain valued rural lifestyles and spaces.

The benefits, meanwhile, would not accrue solely to humans. Cultivated meat could create a *place* for those animals who the abolitionist would 'liberate' through extinction. (I do not believe, recall, that animal liberation *cannot* mean extinction—but I do hold that this would be *surprising*.) Cultivated meat offers a chance for a different relationship with (historically) farmed animals—a chance for us to live with them as equals.

Consequently, cultivated meat could provide many of the goods associated with meat *consumption* while nonetheless being a part of an animal-rights-respecting food system (as argued in Chapter 4). *And* it could be important in the realization of many *other* goods—or so I will argue. To repeat, it is impossible to demonstrate that a particular food system is *all things considered* preferable to another in a work like this. Nonetheless, we will see that there are many reasons to prefer a system with cultivated meat to a plant-based system. We will also see that many of the putative advantages of a system utilizing slaughter-based meat over a system utilizing cultivated meat fade if we produce cultivated meat in ways sufficiently sensitive to the values that people place in meat and meat production.

This exploration also allows us to conclude some unfinished business from the previous chapter. Specifically, it allows us to say more about the acquisition of animal cells—cultivated meat's 'seeds'.

The chapter will advance as follows. First, drawing upon existing critiques of cultivated meat, I identify three key virtues that an ideal meat supply chain would realize. These concern animals, power, and the many values of meat. Second, I look to two existing utopian visions for cultivated meat production—the 'pig in the backyard' and 'mail-order cells'—arguing that they could be a part of an animal-rights respecting food system. Not only that, but they could realize some of the virtues identified. Next, I complement these with a third model, based on the idea of animals as workers; this, I argue, realizes yet more of the virtues. Finally, I consider some objections to the system sketched.

Three virtues

Critics fear that a food system incorporating cultivated meat will fail to instantiate certain virtues present in other food systems. Drawing from these critiques, we can identify three (sets of) virtues that a best-case cultivated meat system would realize.

Virtue one: Respect for animals

As explored in Chapter 4, vegan critics of cultivated meat challenge the idea that its production, promotion, or consumption can be compatible with animal rights. I have argued that, in principle, cultivated meat is consistent with animal rights, *provided* the cell-sourcing method is respectful. But I am yet to identify such a method.

As we saw in Chapter 1, however, it may not be enough to simply ensure that our food system respects animals' rights if we seek animal liberation. It would be curious indeed if animal liberation, animal rights, or justice for animals necessitated *the end* of (many) animals—and critics contend that this is what a food system relying on cultivated meat would mean (e.g. Fairlie 2010, chapter 15).

That said, animal liberation surely means *fewer* (historically farmed) animals. We create the animals most vulnerable to humans in the food system so that we can kill them (or exploit, then kill, them). And they suffer, in part, *because* there are so many of them. And animal liberation would also likely mean that certain *breeds* were no more. Pastoralists have created animals who, because of their biological makeup, live lives of suffering. It would be better *for them*—the animals—if they did not exist. (More on these questions in Chapter 6.) Maybe something similar is true for many animals used in industrial agriculture. But the same is *not* true for *some* of the animals today farmed for meat, who *could* live happy lives.

It would be strange if, once we finally recognized their rights, animals who could live good lives were to (all but) die out. But, to recall the arguments of Chapter 1, this is what existing animal-rights theory apparently requires. Old animal rights is 'abolitionist', actively calling for the end of (all) domesticated animals. New animal rights, typified by Donaldson and Kymlicka's zoopolitics, fares little better. Even while criticizing abolitionism, Donaldson and Kymlicka offer little by way of *a place* for historically farmed animals in their system. Perhaps the occasional person will choose to live with historically farmed animals (Donaldson and Kymlicka 2011, 139), or even form an 'intentional community' with them (Donaldson and Kymlicka 2015). But the closest the authors come to finding a home for large numbers of these animals is when they talk about these animals working, or otherwise contributing to the mixed-species society of which they are members (e.g. Donaldson and Kymlicka 2011, 139–142). This is not to say that animals *must* work (Donaldson and Kymlicka 2020)—but it is to recognize that there can be value in working for the animals (Cochrane 2020), and that framing animals as sources of labour can be a part of recognizing their full societal membership (Wayne 2013), and even a route to animal liberation (Kymlicka 2017).

This is our first virtue. An ideal cultivated meat industry will respect animal rights when acquiring cells, but also find a place for animals 'liberated' by a shift to animal rights.

Virtue two: Power

Critics of cultivated meat contend that cellular agriculture involves a consolidation of power in the food system (e.g. Miller 2012). The technology and materials needed to create cultivated meat, contend its critics, are not available to just anyone. They are available only to those with capital, and may be protected by intellectual property laws. On the other hand, small-scale farming—in theory—is accessible to anyone with land (or access to common land) and the small amount of money needed to acquire tools and some animals or seeds.

For food ethicists (and especially advocates of food justice), a concentration of power is worrying. As individuals, humans have a right to food. Without it, they die. Meanwhile, communities cannot flourish while lacking food sovereignty (control over their own food systems) or food security (access to sufficient food).[1] If cultivated meat replaced slaughter-based meat *and* a small number had control over cultivated meat, then it is true that many communities would likely lack food sovereignty, and individuals' and communities' food security would be at risk.

The imposition of cultivated meat from the outside (that is, those outside of a community pushing cultivated meat into that community) could also erase the food cultures of the community in question. Gone are specific foods, production methods, and customs—replaced, worry critics, with the hegemonic foods. Expect hamburgers, hot dogs, and mince; expect pork, beef, and chicken; expect processed foods. Do not expect foods suited to the palates, kitchens, or customs of the people.

But we must also be careful not to go too far in the *other* direction. A decentralized system must still be a *system*. The production of cultivated meat cannot be a mere hobby; cultivated meat cannot be a novelty or specialist product. To be glib, we are trying to feed the world, not just foodies.

Thus, we have our second virtue. A system incorporating cultivated meat—although it must still be a *system*—will be worse if it consolidates power in the hands of a few, putting marginalized people(s) at risk. A plant-based system (a critic will press) would be preferable. And those who respect human, but not animal, rights would object, too: for them, the end of animal slaughter is a poor trade for the consolidation of the power over the food system.

Virtue three: Values of meat

In Chapter 4, I pointed to the benefits of a system utilizing cellular agriculture (over a plant-based system) relating to access to favoured foods. People value animal products for (among others) personal, social, cultural, aesthetic, and religious reasons. But, as I detailed in Chapter 1, the *production*, not just consumption, of meat involves values that we risk losing in moves away from slaughter-based agriculture.

For many, there is something important about life on the farm. *Agrarian philosophies of agriculture* hold that there is something distinctive, special, *uniquely valuable* about farming. These philosophies champion the preservation of the 'family farm' in particular (Thompson 2018, 53). Agrarians (i.e. advocates of agrarian philosophies) differ on *what* is important about

---

[1] These are traditional concerns of food justice (narrowly construed). Note that *community* is deliberately vague, here; we could be talking about a little-contacted tribe in Brazil or a disadvantaged racialized group in Boston.

farming. In Chapter 1, I drew upon Aldo Leopold to illustrate valued relationships to the land, and I drew upon Jocelyne Porcher to illustrate valued relationships with animals. Such ideas could motivate agrarianism.

Agrarianism's significance should be clear. Agrarians join animal advocates in condemning industrial animal agriculture, but they reject the *complete* abolition of animal agriculture. Something distinctively valuable, they say, is lost with the end of small pastoral farms. Now, these claims cannot mean we should ignore animal rights. Just as agrarian claims about value could not justify worker abuse on farms, so they cannot justify animal abuse on farms. But they *are* arguments that we need to take seriously. Sincerely, and in good faith, many believe that the good life involves working closely with the land, animals, and/or food in ways only a small, pastoral farm provides. Sincerely, they believe these farms are worth preserving—the world would be worse without them. We can take these values seriously by asking whether we could preserve these farms in a just system. In producing cultivated meat, can we respect animal rights *and* agrarian values?

It is not just agrarians who value animal farming. Again echoing Chapter 1, there is good work associated with pastoralism—on or around farms, and in associated industries. Valued rural ways of life (customs, meals, festivals …) are built around farming. Even urbanites value visions of the countryside. Agricultural animals are part of the rural visions we value (as environmentalists, naturalists, nationalists, artists, hobbyists, tourists …). And the animals and farmers determine which plants thrive, the composition of the soil, and the lay of the land itself. If we lose the farm, much that is valued is lost, too.

Family farms are also important for foodies. Being close (for locavores, *physically* close) to food sources can be important. So, for example, advocates of 'slow food' object to reductionist understandings of food ostensibly presented by cultivated meat, eliminating 'culture, tradition, skill, difference, and context' found in small-scale farming, as well as the way that removing the small farm 'alienates people from their food, hides the production process, and culturally (and spiritually) impoverishes food and agriculture' (Sandler 2015, 143). The family farm ostensibly allows for these things.

Before concluding this subsection, it is worth acknowledging another concern: Authenticity (Wurgaft 2020, chapter 11). (Compare the discussion of plant-based meat's unnaturalness in Chapter 3.) The idea of authenticity in food is notoriously slippery, but the rough idea is that certain foods are not what they purport to be: Pizza Hut is not *really* serving pizza, which is authentically Italian; some farmers are not *really* growing wheat, as authentic wheat has not been genetically modified; busy professionals are not *really* cooking, as authentic food preparation needs more than a microwave. The idea is that, at best, cultivated meat is an imitation of authentic meat.

It is difficult to pinpoint authenticity, and why it is important. It is, regardless, valued. This may seem to be a difficult challenge for cultivated meat. I note, however, two responses. First, as discussed later, certain forms of cultivated meat production might be 'more' authentic than others, and there are ways that cultivated meat could retain ties to traditional foods.

Second, I am not proposing that the state forces people to eat cultivated meat. If a concern for authenticity means that cultivated meat is not accessible to some people, so be it—they can eat authentic plant foods (or other authentic, but rights-respecting, animal products). The value some people place in authenticity cannot trump animal rights, just as it cannot trump human rights. We would not give weight to the opinion of someone who insisted that their meal was only *authentic* if prepared by underpaid workers, and who therefore refused to buy from restaurants with fairly paid staff.

The best food system would preserve not just the values associated with access to meat, but the values associated with the small family farm as the source of that meat. This is the third of our virtues.

Two existing visions

There is a large and growing academic literature on cultivated meat. Nonetheless, there are few worked-out visions of how cultivated meat might fit into an animal-rights-respecting food system. Two jump from the literature. We can call these *the pig in the backyard* and *mail-order cells*. We can explore these as visions of the ideal. It is my argument that both could be a part of the zoopolis's food system, but that both are, in important senses, lacking. I thus offer a third possibility, complementing the first two, drawing on the idea of animals as workers.[2] A combination of these three visions, I argue, offers a positive vision of cultivated meat's home in the zoopolis.

The pig in the backyard

The pig in the backyard (PIB) is the thought

> that in the future we might all have a pig in our backyard or in our local community, from which some stem cells are taken every few weeks in order to grow our own meat, either in a machine on our kitchen sink or in a local factory. It is an idea that in

---

[2] Although distinct from any existing proposals, this third model has affinities with the sanctuary model offered as a non-ideal theory by Jan Dutkiewicz and Elan Abrell (2021).

some form or another often turns up in conversations on cultured meat. It typically takes the form of pigs or cows in urban farms or backyards, held as pets and serving as donors of muscle stem cells.

(Weele and Driessen 2013, 655)

PIB emerged from a workshop of academics interested in cultivated meat production, and gained clear support. Cor van der Weele and Clemens Driessen, who led the session, said the enthusiasm for PIB probably had

> something to do with the unexpected combination of many good things that have always seemed incompatible. Here, all of a sudden, we get a glimpse of a possible world in which we can have it all: meat, the end of animal suffering, the company of animals and simple technology close to our homes. The pig in the backyard or in the community, that is a pet and a cell donor for cultured meat at the same time, creates the possibility of sharing the world with animals in sustainable as well as conscientious ways, while we do not have to give up eating meat. The prospect eliminates the suffering caused by intensive farming, but not by replacing it with an abolitionist world in which urban vegans are completely separated from nature and from animals, pictured by [Simon Fairlie (2010, chapter 15)] in such bleak terms. On the contrary, in this possible world relations between humans and animals improve spectacularly. In an idealized version, it would even improve upon practices such as chickens kept in a backyard for eggs or honeybees kept on an urban rooftop, as these modes of food production require more significant interfering with the behavior and circumstances of the animals. The feeling that this prospect is almost too good to be true may help to explain the special atmosphere.
>
> (Weele and Driessen 2013, 655)

Let us test this vision against our virtues.

At first glance, PIB does well on respect for animals. (Putting aside, for now, the question of what it means to '[take] some stem cells ... every few weeks'.) It finds a place for historically farmed animals—they live as 'pets', in the 'backyard' or 'community'. And, if they are 'pets' or companions (beloved family members) there is every hope that they will have happy lives. Certainly, they will have happy lives compared to the lives of pigs used for slaughter-based meat. Even those on the most 'humane' farm, unlike companions, face a gruesome, early death.

The idea may sound speculative, but cellular agriculturalists have put something like it into practice. For example, a proof-of-concept promotional video produced by cultured meat company Eat Just, Inc. [the producer of the first commercially available cultivated meat] shows a group of humans dining on cultured chicken nuggets at a backyard picnic table while Ian, the chicken whose cultured meat they are consuming, walks around in the grass at their feet. A voiceover narration from one of the diners explains: 'It was an out of body experience to sit there and eat a chicken, but have the chicken that you're eating running around in front of you. You don't imagine doing something like that, but then you have this realization that we've figured out how life really works, and now we don't need to cause death in order to create food. And we're going to have to do it if we want to continue living on this planet'.

(Dutkiewicz and Abrell 2021, 6)

One of the problems with PIB is that it seems uncommercializable. If it was commercialized, it would be wrong to call the eponymous pigs 'pets'. People live with companions because they value the companion and her company. This gives them clear motivation to care for them. If they are benefiting from the pig's presence in assorted ways—including having access to cells for bacon—that need not be an issue (cf. Fischer and Milburn 2019). But if they keep a pig for *profit*, the relationship seems to have shifted.

While we could compare the relationship one has with a typical companion to that one has with a child (see, e.g., Palmer 2010, 94)—we love both for their own sakes—the relationship with the pig kept as a money-maker is different. The appropriate comparison now seems to be with an employee. But, of course, the state (appropriately!) affords rights to employees that it does not to children. While there is no reason to believe that 'pets' could not be the source of cells for cultivated meat production—just as we can ask children to contribute to the running of a household, as far as they are able—to plan that they would be the *primary* source of such cells seemingly misconstrues the relationship we have with them. This is a problem for PIB.

At first glance, PIB does well on the centralization of the food system. Assuming easy access to the machines present in the 'kitchen' or 'community factory' of PIB, the vision presented is *radically* decentralized—corporations/states would have little control, just as they have little control over what gardeners grow in their vegetable patches.

There are, though, two worries. The first is that PIB relies upon as-yet unavailable technology. Maybe there could be affordable technology developed that allows typical consumers to produce meat in the way imagined in PIB—but we do not have it yet (Dutkiewicz and Abrell 2021, 6).

The second is that PIB goes too far in the *other* direction. Could PIB be the basis of a food *system*? Could it provide a reliable, steady supply of cells?

Could it provide the *variety* of cells desired? There is something appealing about keeping pigs then creating bacon using nifty gadgets. But what about those people who are not part of the rural or suburban neighbourhoods where there is space for pigs? There is no room for pigs (let alone cattle!) in inner-city neighbourhoods. People who rent or frequently travel cannot look after pigs. People already working multiple jobs and caring for human dependants do not have time for pigs. If these people are lucky, maybe (*maybe*) they can get meat from a neighbour or a 'community' pig—much as they may be able to get fresh vegetables from a neighbour or community garden. But whether they can access cultivated meat in other places—restaurants, supermarkets, work (school, prison …) cafeterias, etc.—remains unclear. PIB may be utopic for a certain kind of middle-class consumer. But it may not be for others.

On the many values of meat, PIB is mixed. It provides a close relationship with animals and local food, so foodies, in principle, could cheer for it. Even those concerned with 'authentic' food could concede that being able to see (hear, smell) the *actual* animal whose meat you are eating—even *while you are eating*—could provide an authentic food experience.

Agrarians, though, will not welcome rethinking agricultural animals as (mere) pets, even if they could welcome many people taking up (a sort of) farming. But the loss of the family farm will be the real blow. And with the loss of the farm will come the loss of animals in the countryside, the change to the rural economy, the change to the landscape, and more. PIB, although compelling for some who *consume* meat, likely is not appealing to those who (currently) *produce* it.

While there is surely room for practices reminiscent of PIB in the zoopolis, it cannot—alone—provide a foundation for a food system that scores well on the virtues identified.

Mail-order cells

Isha Datar, the Executive Director of the cellular-agriculture charity New Harvest (and the person who coined the term *cellular agriculture*), also has a vision of locally grown cultivated meat. Unlike PIB, her model—which we can call *mail-order cells*, or MOC—does not place animals at its centre.

> Imagine producing meat at home without killing animals. With a few cells and a keg, the process could be no more complicated than brewing your own beer or pickling vegetables. … [When it comes to producing good food, e]xperimentation

> will be key. But the first hurdle often faced by enthusiasts is obtaining cells to start the process. At the moment, muscle stem cells are most easily obtained from fresh meat at a slaughterhouse or from live animals—preferably young ones since their stem cells are more plentiful. But harvesting them is hard work.... Datar hopes to change that by making cell lines available for order from lab supply catalogues or by linking up researchers so those with cultures can share them with others, much as people share sourdough starters to make bread. For Datar, 'it would be like open-source software. The cells are the code'.
>
> (Ceurstemont 2017)

MOC is an image of a group of home biohackers growing meat from cells, just as home bakers grow their own sourdough starters. Animals are noticeably, deliberately, absent; animals are an inconvenience.

Cells need to come from somewhere, of course. But in MOC, as an ideal, animal involvement will be minimal. Perhaps cells will come from (close to) immortal cell cultures, which ultimately owe their origins to living, breathing animals—but can be sustained without them. The image of cells from a single animal going on to feed (literally) billions of people is an image drawn upon in the more hopeful visions of cultivated meat's potential (see, e.g., Stephens 2013). These optimistic visions are of a future in which we 'make animals obsolete' (Abrell 2021, 41).[3]

Admittedly, Datar is not engaging in ideal theory when she proposes MOC. But others *do* propose this as an ideal. Jan Dutkiewicz and Elan Abrell, although focused on a *non-ideal* vision of cell-sourcing, suggest it is feasible that 'donor animals are a necessary transitional step to a cellular agricultural industry that runs on immortalized cells' (2021, 7), while envisioning that 'as soon as animal cell lines can be effectively immortalized', the use of animals 'would be phased out' (2021, 11).

Doing away with animals might draw upon a range of technologies, many of which do not play a current part (to my knowledge) in cellular agriculture. Crucially, cellular agriculturalists do not (currently) have access to suitable 'immortal' cell-lines for mammals or birds, meaning that 'donor' animals will indeed be necessary for the foreseeable future (Dutkiewicz and Abrell 2021, 5).

Nonetheless, the positives of removing animals from the picture are many. Let me give three sets of advantages. First: No animals means no animal

---

[3] Indeed, maybe cells scientists could create cells in the lab, based on (say) plant cells. Cell biologists have been able to create something like cells for years, and can relatively easily chop and change 'natural' cells into something different. There is no reason that cultivated meat need resemble the meat of any animal who has ever lived.

suffering and death. Second: Removing animals unshackles cellular agriculturalists from the dozen-or-so animals routinely eaten in the west, making their options almost limitless. Third: No animals brings health advantages. For example, if we are collectively not interacting with farmed animals, the risk of zoonoses would be lower.

MOC is, in a sense, impeccable on our animal virtue in that it does not involve the violation of any animal rights. But we can raise questions about animals' *absence*, too. MOC represents, perhaps, the realization of the abolitionist dystopia envisioned by cultivated meat's critics, and, from an animal-rights perspective, it fails to find a 'place' for liberated animals.

MOC is compelling on the centralization of power because of its focus on open-source culture. On the other hand, if particular corporations controlled the catalogues from which cells were sourced, power would be centralized. As with PIB, we can raise questions about access to the education and hardware necessary to produce cultivated meat at home. But, unlike PIB, MOC has a clear route to being scaled up. While making meat may start out as a pursuit of (the equivalent of) home brewers, production could in time be taken up by small and large businesses—as well as schools and other public institutions—ensuring a steady supply. MOC could ground a food *system* in a way that PIB could not.

MOC, however, does not fare well on the many values of meat. It certainly allows for (borrowing agrarian language) skills and knowledge, but it does not need farmers' *existing* skills and knowledge. MOC would mean the end of the (pastoral) family farm, replacing it with a factory—a dystopic vision for the agrarian. Meanwhile, MOC represents, perhaps, the ultimate (to coin a term) artificialized food, the ultimate inauthentic food. This will be, for many, distasteful.

For all this, though, MOC could form a secure, safe foundation for a food system incorporating meat—and one that need not entail the monopolization of food production. It could not, however, form the *entire* basis of meat production—or not without the sacrifice of many people's values. Better to complement it with further sources of cells.

A third vision: Animal labour rights

My proposed addition to PIB and MOC is, in short, that animals remain the source of cells, but that, rather than pushing them into the backyard, we keep them on the farm. These animals are entitled to a full suite of rights *qua* animals. But they are also entitled to more. As farms are not, and have never been, animal-friendly places, I propose that we protect the animals in question with *workers' rights*.

First, I must say something about animal worker rights.[4] Once I have done this, we will be able to sketch this proposed 'farm'—and how it would fit into the food system.

Animal labour rights?

A central concern of new animal rights has been exploring the positions that animals inhabit relative to humans (and the mixed-species community), and the rights to which these animals are entitled *qua* beings in those positions. So, for example, certain animals may be citizens (Donaldson and Kymlicka 2011), property holders (Hadley 2015), or refugees (Derham and Mathews 2020).

For humans, each of these statuses brings rights *complementing* basic human rights. Citizens have a right to political representation in a way non-citizens do not—but I can kill neither. Property holders have a right to use objects in a way non-owners do not—but I can torture neither. And refugees have a right to claim asylum when other migrants do not—but I can enslave neither. So, too, in the animal case. While all (sentient) animals have certain basic rights, animal citizens, animal property-holders, and animal refugees have rights *on top of* those basic rights.

One position that has attracted particular attention has been the animal worker (see, e.g., Blattner, Coulter, and Kymlicka 2020). In the human case, a long history of politics, scholarship, and struggle has secured legal rights for individuals *qua* workers. For example, as a university employee, I have rights not to be fired without good reason; to pay; to belong to a union. Those rights did not apply (in the same way) when I was an undergraduate. Nor would they apply if I became unemployed, but continued to attend university colloquia. And my possession of these rights is unrelated to my possession of basic human rights.

We should not assume that we can simply apply all the rights possessed by humans *qua* workers to animals *qua* workers, as human workers and animal workers may have different interests. For example, even if human workers should have a right to access continuing education, it is not obvious that animal workers should. This does not mean that animals are less worthy. Equally, I have no right to time off when pregnant, even while some colleagues do.

---

[4] For the purposes of this chapter, I am treating *labourer* and *worker* as synonymous, and thus treating *labour rights* and *workers' rights* as synonymous.

But this is not because I am worth less than my colleagues. It is because I lack a womb, so have no interest in this right.

Others have done the groundwork of identifying the rights that would benefit animal workers. Alasdair Cochrane (2016, 27) identifies the following five:

1. The right to representation by a labour union.
2. The right to a decent standard of remuneration.
3. The right to healthy and safe working conditions.
4. The right to rest and leisure.
5. The right to a decent retirement.

José Luis Rey Pérez (2018)[5] identifies the following six:

1. The right to recognition, including a salary, food, and care.
2. The right to a limited working day, including regular days off, to allow time for rest and play.
3. The right to holidays.
4. The right to safe and hygienic working conditions, allowing animals space.
5. The right to social protections, including retirement.
6. The right to participate in workplace decisions.

The two lists have significant overlap. Cochrane's right to rest and leisure chimes with Pérez's rights to limited working days and days off; both posit rights to remuneration, safe working environments, and retirement. Meanwhile, the right to membership in a union, as proposed by Cochrane, provides an important route to the realization of some of Pérez's rights. As such, I take it that Cochrane and Pérez mostly agree.

Something included in neither Cochrane nor Pérez's lists, although endorsed by many others writing about animals as workers (e.g. Blattner 2020), is a right to choose one's work. Thus, based on the existing literature, we can identify the following list of rights for animal workers:

1. The right to representation by a labour union, including the right to participation (via the union) in workplace decision-making.
2. The right to a decent standard of remuneration.
3. The right to healthy, hygienic, and safe working conditions.

---

[5] Pérez 2018 is only available in Spanish. I base this list on Pérez's 2019 presentation 'Labour rights for animals?'

4. The right to time off work including limits to working days and time off work to engage in rest, leisure, and play.
5. The right to retirement and aid in case of unemployment.
6. The right to a choice in one's work.

I note three things about this list. First, it is not final. Perhaps animal workers should have some other rights. Perhaps they should not have some on this list. Perhaps we need to adapt the list to different kinds of work, and different animals.[6] But I take it that, as it is drawn from the existing work on animals as labourers, it provides a useful starting point.

Second, some of these rights require fleshing out with particularities, even if they are clear enough in the abstract. A safe working environment is easy enough to envisage, although different workplaces and animals will have different requirements. I am not going to get into details, as there are any number of animals, with any number of needs, who may be farm labourers for the purposes of cultivated meat production. For instance, ethicists could write whole chapters about cattle horns.[7] We can only answer questions like that with attention to particularities. Indeed, we will have to work out answers to some questions in practice, with negotiation between animals, legislation, farmers, policymakers, experts, and the public. Consequently, there is much I cannot answer here. However, I consider egg-laying chickens in Chapter 6. That discussion gives an idea of what details may look like in *one* case, and could provide a model to assist in others.

Third, some of these rights may be difficult to envisage even in the abstract. What, precisely, would a union for animal labourers look like, given that they cannot participate in the election of representatives? What does it mean to *pay* animal workers, given that animals have no understanding of money? I do not want to go too far into these questions in this chapter, which focuses on cultivated meat rather than animal workers. I delve into them more deeply in Chapter 6.

Down on the farm

Let me, then, return to my practical proposal. The idea is that cellular agriculturalists source cells from animals living on farms, but that these animals

---

[6] Again, this happens in the human case. Child workers can have different rights to adult workers. Meanwhile, certain kinds of work can carry with them different kinds of rights—sex workers and academics need different protections.

[7] Should they remain? Are there humane ways to remove them? Should farmers utilize only polled breeds? May genetic technologies be deployed to create polled strains of horned breeds? And so on.

are not the mere property of the farmer. Instead, they are protected by a suite of workers' rights, more like contemporary farmhands than contemporary farmed animals. And they are not 'donors' of cells. Instead, they are remunerated for their services.

The animals would live in a space that would resemble a contemporary small-scale farm or animal sanctuary (Dutkiewicz and Abrell 2021). They would be free, for the most part, to live out their lives as they saw fit within this space. Thus, a farm of this sort would not raise concerns for some of our workers' rights: the animals are in a healthy, hygienic, and safe environment; they have plenty of time for play, leisure, and rest; and worries about choice in work are minimized by the fact that we are asking them to do little beyond living their life. Union membership and social support protect the animals, and they receive remuneration.

Periodically, these animals would have cells taken from them for the purposes of growing cultivated meat. These 'harvests' would be uncommon, given the productivity of cultivated meat technology, and only younger animals would be used—not only are younger animals better sources of stem cells (Ceurstemont 2017), but older animals would be allowed to enjoy their retirement.

Precisely how this 'harvest' would take place is, of course, crucial, although details may vary from animal-to-animal. In short, farmers must take material from animals, and harvest suitable cells from this material. Meat contains different cells (muscle cells, fat cells, etc.), and *other* kinds of cells (such as assorted stem cells) can grow into these 'meat' cells. Cellular agriculturalists can use various methods to acquire stem cells. Which are suitable (and effective) for which kinds of meat production raise many scientific questions (see, e.g., Melzener et al. 2021). Consequently, it is inappropriate to decree the cell-sourcing methods that farmers should use. For example, there is ongoing work exploring which parts of animals' bodies provide the best cells in the largest numbers, and this may vary from species-to-species, breed-to-breed, age-to-age—even animal-to-animal. What is important, instead, is that the approaches taken combine feasibility with ethical acceptability.

Let us consider some examples. Cellular agriculturalists producing chicken could source stem cells from eggs,[8] from feathers,[9] from slaughtered chickens, or from biopsies. As well as scientific questions, each raises its own ethical issues. The sourcing of cells from slaughtered chickens is not compatible with animal rights. The sourcing of cells from eggs raises questions about

---

[8] An approach taken by SuperMeat, which produced the cultivated meat available in Israel.
[9] An approach taken by Eat Just, which sold cultivated meat in Singapore. Recall Ian the chicken.

the sourcing of the eggs (discussed in Chapter 6) but would, in principle, be acceptable. Similar is true of feathers.

When it comes to mammals, options are more limited; the choice, on the face of it, is between biopsy methods. Needle biopsies are one possibility, and involve no more pain than the taking of blood. (Incision biopsies involve taking more of the animal's body; 'discomfort will likely be greater [than with needle biopsies], although still not substantial' (Melzener et al. 2021, 10).) The use of pain relief is, of course, possible. And so, with creative thought, is the possibility of sourcing stem cells without interfering with animals' bodies. Umbilical cords, for example, can be the source of stem cells, and this has seen experimental success. Early cellular agriculturalists used pig umbilical cords (Stephens 2010, 397), and this has the potential to be an ethical (and productive) alternative to biopsies (Stephens 2013, 169–170).

But let us focus on biopsies. As indicated, there is no reason that a biopsy could not be painless (Bhat et al. 2019, 1202). Biopsies would be rare, and need not involve any coercive behaviour on the part of the farmers. Farmers could take biopsies as part of routine human–animal contact. Or farmers could encourage animals to participate in their own time, at their own speed, perhaps in exchange for a favoured food (or similar). Ensuring that animals were neither harmed nor particularly inconvenienced by biopsies would be crucial to ensure that farmers respect animals' right to choose their work. If biopsies neither harm nor inconvenience animals, we can sincerely say that the animals have chosen the shape of their life and work; if the biopsies do harm or inconvenience the animals, however, we need some fuller story about the animals' choice, and that may be tricky to provide.[10]

Farmers (or other professionals) can use these cells to create a cell line, and this provides the source material for cellular agriculturalists. Incidentally, the animal worker model is compatible with PIB's vision of the kitchen appliance or community-owned factory, and with MOC's catalogue of cells. However, it can effectively fill the gaps left by both PIB and MOC by preserving the farm.

This conception of animal workers provides a home for newly liberated animals, but, more importantly, it provides a *place* for them—and protections. The state provides oversight of the relationship these animals have with their caregivers, and the political community, of which these animals are a part, respects their rights. This is not a temporary fix while biologists develop

---

[10] It is one thing to observe that one dog prefers spending time with the animals she guards while another prefers to spend time around humans for whom he provides a therapeutic benefit (see Donaldson and Kymlicka 2011, 139–140). It is another to say that some animals choose to provide material for a high-tech industry they cannot comprehend.

immortal cell lines (as in Dutkiewicz and Abrell 2021)—this is a proposal for a lasting relationship, a place for liberated animals for the foreseeable future.

This vision of animal workers also fares well relative to concerns about the centralization of power. The system's foundations are the (human and animal) workers on farms, and neither corporations nor states. There is scope for farmers to sell their product directly to consumers (either by producing meat on-site, or by selling cells to home-growers). But this is no mere hobby. Again, it has the potential to form the basis of a *system*, either complementing or underpinning an MOC model.

The key complement that this proposed model can offer to *both* PIB and MOC, however, is that it preserves an institution closely resembling the family farm. This preserves good work—replacing inhumane jobs with humane jobs (Coulter and Milburn 2022), a central goal for 'food tech justice' (Broad 2019; Abrell 2021, 46–47).

It is also of paramount importance to agrarians, for whom the family farm provides a uniquely valuable way of relating to the world. It is true that the model does not preserve *some* aspects of the family farm. There is no animal slaughter, castration without anaesthetic, or 'breeding' on a stanchion ('rape rack'). But I contend, based on their own words, that these are not the things agrarians deem valuable. Indeed, it would be hard to imagine why these unpleasant (disturbing?) parts of the job would be the parts worth preserving. Better to preserve the good parts and lose the ugly parts. Perhaps, indeed, agrarians could come to see a greater value in the respectful family farm than in the farm replete with death and suffering (compare Delon 2020).

The importance of the family farm for agrarians and foodies has already been explored. Indeed, the connection that people can have to their food is arguably *stronger* on this model than in (for example) the contemporary locavore's lifestyle. Foodies could visit and sit with the *actual* animal whose meat they eat *while they are eating it*. Think of Ian the chicken. It is hard to imagine a more authentic way to—borrowing a locavore phrase—meet one's meat.

Advocates of PIB described the model as a chance to 'have it all' (Weele and Driessen 2013, 655). I contend that PIB offers a chance for a particular kind of (suburban?) consumer to 'have it all'—it would mean a loss for many in rural environments, with ties to farms. In response, I propose that PIB and MOC be complemented with an animal-respecting form of animal agriculture. An animal worker model is a chance to *genuinely* have it all. Animals' rights are respected; meat is plentiful for those who want or need it; power is decentralized; the (for some) uniquely valuable institution of the family farm is preserved; liberated animals have a home and place. These are, in my view, worthy goals for the food system of the zoopolis.

A miscellany of objections

I have offered an outline of my proposal. Recall that I am *not* saying that animal workers are the only way cellular agriculturalists could produce cultivated meat in the zoopolis. I am saying that an animal worker model complements PIB and MOC. I provide indicative details of how some parts of this model would work in Chapter 6. It is worth closing this chapter, however, with a consideration of possible objections.

Corpse farming

If, as proposed, we can keep animals on farms to produce meat, why do we need cellular agriculture?

Some advocates of the idea of animal workers propose that farmers could keep and slaughter animal workers (e.g. Porcher 2017). This is inconsistent with *both* animals' basic rights and their rights *qua* workers. Rather than animal workers being a concept with emancipatory potential (see Blattner, Coulter, and Kymlicka 2020, passim), this is apologism for animal abuse (Delon 2020).

A more interesting proposal is (what we can call) corpse farming. The idea is that while it violates animals' rights to kill them and make them suffer, it does not violate their rights to keep them on farms and eat their corpses when they die (Zamir 2008, 49; Cochrane 2012, 87–88). A related proposal, from Jeff McMahan (2008; cf. Lamey 2019, 194–197), suggests that we genetically modify animals to die naturally and painlessly when young. The animals in question have no obvious complaint,[11] and humans have access to meat. And, we can add, the valuable institution of the farm is preserved.

Although I accept that both keeping animals on farms and eating meat are (in principle) acceptable, I reject corpse farming. First, the proposal offers farmed animals no chance of retirement. Animals farmed for their corpses would be workers, and thus entitled to workers' rights. But this includes a right to retirement, and there is no possibility of retirement for these animals. Corpse-farming, then, seems to violate the rights of animals *qua* workers.

I have a second, separate, concern. It is one thing to say that it is permissible to eat meat. It is another to say that it is permissible to butcher a given animal's corpse. Animals on farms warrant protections as members of our community, and this, we might hold, includes affording to their bodies

---

[11] We could not object that if the animals were unmodified, they would have lived longer. If they were unmodified, they would be different animals.

the respect that we extend to community members' bodies (Donaldson and Kymlicka 2011, 151; Milburn 2020). Contemporary liberal societies do not see butchering a corpse for food as consistent with respecting said corpse. Given this, and assuming that society members' corpses do warrant respect, it would be wrong to butcher the bodies of animals who live on farms. So not only does corpse farming fail to respect animals *qua* workers, it fails to respect them *qua* community members.

For all that, corpse farming remains a possibility to take seriously. Perhaps these problems can be overcome, and there would be a place for corpse-farming in the zoopolis. For example, Jeff Sebo (private correspondence) suggests that my claims about genetically modified animals fall short: Claims about retirement may not apply to working animals with very different life cycles to those with which we are familiar, and claims about corpse respect are contingent on current attitudes. I am open to the former contention, and freely accept the latter.

This, though, leaves further problems. Key among these[12] are the decidedly non-ideal questions of, first, whether we *could* develop animals that painlessly died leaving ample, desirable meat, and, second, whether said development could respect animals' rights. But what should be clear is that allowing the 'farming' of animals for cells does not *necessarily* open the door to corpse farming.

The model is not generalizable

This model has been designed thinking about certain animals in certain contexts. Namely, I have developed it thinking about the sourcing of cells from pigs, cows, sheep, chickens, ducks, turkeys, and so on in western democracies with strong agrarian traditions. It does not necessarily work when it comes to sourcing cells from other animals whose meat some people may want to consume, and it does not necessarily work when it comes to communities where workers' rights—if they are recognized at all—play out differently.

I concede the first point. Yes, pigs and chickens can be 'farmed', and no, whales and chimps cannot. Nonetheless, people *do* eat whales and chimps, and might value continuing to do so. We will need different models to source cells from whales and chimps. We will need different models, too, if we want to step beyond those animals currently eaten. But this is one of the values of a mixed system. Even if some animals are not well-suited to an animal worker

[12] Sebo suggests that 'the problem is with our selfish priorities when engineering these animals' (private correspondence). I struggle to pin this down. Selfishness is unvirtuous, but I am not convinced it is unjust.

model, one of the other models—or an entirely new model—may work. For example, cellular agriculturalists can produce fish meat with immortal cell lines, meaning there is no 'need' for fish farms (Dutkiewicz and Abrell 2021, 5). There may be room for them, though, depending on, first, whether aquaculturalists could 'farm' fish in a way consistent with their rights, and, second, whether there was a demand for aquacultural institutions.

The second point, too, has merit. It would be inappropriate to simply *assume* that we could export an agrarian-inspired model of workers' rights developed within a western liberal democracy to (say) nomadic herders among the Maasai of East Africa, the Sámi of the Nordic countries, or Turkic peoples of Asia. Perhaps animal labour rights *would* work in these environments. Or perhaps theorists could develop rights-respecting models better suited to these socio-political cultures.[13] Or—although this raises the spectre of cultural imperialism—perhaps non-liberal communities should liberalize, moving towards a culture in which workers' rights (human rights, animal rights, justice ...) are more acceptable. For the most part, I leave these questions aside. They raise bigger issues than I can satisfactorily answer here.

However, I suggest that this model of workers' rights for animals *could* work well enough in much of the western and non-western world, specifically for those terrestrial vertebrates most routinely farmed—cows, pigs, sheep, goats, chickens, turkeys, rabbits, ducks, geese, and so on.

Conflict with other values

This model of workers' rights for animals may be able to provide a place for and protect animals while preserving many of the values associated with the small pastoral farm. However, it might seem to conflict with *other* values that could be associated with plant-based food systems—lowered climate impact, lowered zoonoses risk, and so on.

In a sense, this is unsurprising. Enduring, pervasive, and fundamental disagreement about what is valuable, what we should preserve, and how we should live is (in free societies) a part of the human condition. There is thus a sense in which we should accept that some people might want to see things done differently, and leave it to individuals to determine how they would like to live, through choices about (for example) careers, hobbies, and purchases. (States, of course, must negotiate this disagreement. More in Chapter 7.) Recall that my proposal here is not for a top-down, state-run

---

[13] I am not committed to the claim that non-liberal states cannot protect human and animal rights, although exploring this fully will take us from the current inquiry.

food system that produces and distributes food to all. (Although perhaps that would be compatible with my arguments; again, more in Chapter 7.) Instead, I am proposing a food system that would be permissible (crucially: in accordance with animal rights), and that would (or, *could*, *might*) be all-things-considered preferable to a plant-based, wholefoods system.

The state can be a crucial actor in determining the shape of the food system—purchases for public institutions, choices about subsidies, and recommendations related to nutrition are three obvious areas in which the state shapes and influences food systems and diets. (Again, more on this in Chapter 7.) But, in democratic states, the state should not support a food system conflicting with the values of the populace—except when doing so is the only way to discharge its duties, such as (say) ensuring that there is enough food to go around.

If there is no sufficient mass of people choosing to work with animals in the way I have suggested and a sufficient mass of people ready to purchase the products of such farms, the system I am envisioning probably wouldn't (*couldn't*, *shouldn't*) prosper. This includes, of course, if the products of animal workers are not available at prices consumers are willing to pay. So, yes, absent sufficient public buy-in, the animal labour model could not work—and there is a sense in which we could simply say, provided the alternative continued to respect animal rights, so be it.

However, I am not convinced that this hard-nosed response is necessary, as a food system with cultivated meat produced by animal workers *could* realize many of the advantages of a vegan food system unrelated to animal rights. The crucial point is that this is not an agricultural model that resembles intensive animal agriculture. The model minimizes the threat of zoonoses, which is associated closely with intensive animal agriculture but not particularly with non-intensive animal agriculture. The meat itself can be healthier (cellular agriculture offers a greater degree of control over the constituents of meat) for those who desire this, while those who prefer 'authentic' meat products[14] can have them—it is their choice to eat unhealthy products if they wish.[15] While there will be greenhouse-gas emissions, those *currently* emitted in pursuit of meat dwarf them. And although we may be talking about many farms, we are talking about relatively few animals on each, as cellular agriculture significantly increases the level of meat produced per animal. (More on this shortly.) Thus, there will be no need for manure lagoons, routine antibiotic

---

[14] Here meaning something like 'meat products with nutritional profiles matching those available in 2010s'.

[15] Compare the discussion of health in Chapter 3.

use, or growing huge amounts of soy (or similar) to feed animals crammed into sheds.

In countering the challenge that the animal worker model has disadvantages relative to a plant-based system, we can also note potential environmental and health benefits of using animal workers, even relative to a plant-based system. (These advantages may or may not apply to a system utilizing plant-based, but not cultivated, meat.) Some people experience 'allergies' to fruit and vegetables, meaning that meat could be a healthier (or more comfortable) dietary choice. A relatively common condition, for example, is oral allergy syndrome, which results in people with hay fever experiencing negative reactions to fresh fruit and vegetables.

Meanwhile, meat can be a nutrient-dense source of food, providing vitamins lacking in unvaried plant-based diets. Thus, meat can be an important source of food for malnourished and food-insecure people, including both individuals lacking easy access to food (in both the developed and developing word) and those who have relatively easy access to food but have eating disorders (or who are picky eaters). Access to meat may thus have real health benefits for some people.

Small-scale farming could have important environmental benefits related to land management, including topsoil restoration. The case for regenerative agriculture makes for some of the most convincing arguments in favour of small-scale pastoral farming over plant-based systems (see, e.g., Katz-Rosene and Martin 2020), and it is a virtue of the animal worker model that it can incorporate these advantages (cf. Dutkiewicz and Abrell 2021, 9).

I do not rest my case for cultivated meat or animal workers on these advantages—I simply raise them to partially offset (or at least complicate) concerns about how the animal worker model clashes with some of the values realized by a plant-based wholefoods system.

How many animals?

One advantage of the system I have proposed is that it provides a place for formerly farmed animals; another is that it allows animals to be present in the countryside, which people value. But these arguments only work if we are still talking about a moderate number of animals. How many farms could there be?

The short answer is that we do not know how many animals we will need for the cultivated meat industry. This depends not just on the industry's technological advancement, but also the demand for meat in the zoopolis. How many humans and non-herbivorous animals will there be? How many will eat cultivated meat? How much cultivated meat will each eat? Preliminary estimates are provided by Dutkiewicz and Abrell:

> Given the nascent state of the industry, it is difficult to estimate the number of necessary donor animals. Low estimates suggest that biopsies taken from one cow could replace 400 cows over her lifetime. However, improved capacity to multiply cell populations leads some researchers to suggest that one biopsy from a cow could replace up to 13 million cattle (Melzener et al., 2021). Regardless, populations of donor animals like cattle would have to be kept above minimum viable breeding populations, which means that about 20,000 donor cattle might be required to completely replace the beef from the global cattle population (Melzener et al., 2021). This would be the case at the very least in the short and medium term until scientific progress allows mammalian cell lines to be immortalized or synthesized [i.e., a move to the MOC model]. However, this might prove impossible or prohibitively expensive, in which case donor animals would be required as long as cellular agriculture exists.
>
> (Dutkiewicz and Abrell 2021, 5)

Another factor is the question of diversification: Animals workers may have roles (careers!) beyond being cell donors. The animals' presence could have other (economic) benefits (Donaldson and Kymlicka 2011, 135–136): Children (and former children) could visit animals to spend time with, and learn about, them; nature reserves and landowners could employ animals for ecosystem regeneration and management; ruminants could keep grass short in parks, around solar panels, and similar; animals of all kinds could produce manure for arable farmers and horticulturalists. Incidentally, these revenue streams are already familiar to small-scale pastoral farmers.

Nonetheless, assuming streamlined technology, we could imagine a small number of farms being able to provide the volume of (say) pork currently eaten, and this seems to threaten the vision offered. But this assumes only a small number of pork products. It is likely that people will still want access to pork from heritage breeds, or from pigs local to them, or similar. It is likely that people will still want to visit farms and buy meat from the pigs they had just seen. (*Literally* the same pigs.)

More than this, people may not merely *want* to have these things; people may consider access to these things part of their understanding of what it means to live well. And if there is demand for these things, there will be room for them in the zoopolis—even if they are not necessary to provide the quantities of pork currently eaten.

On top of this, there is room for a conversation about whether the state should be providing direct support for farmers working with animals to produce cells for cellular agriculture. Let us not forget that pastoral farmers in many states are *already* in receipt of considerable subsidies. And if the existence of these farms is important for many people's view of the countryside, for many people's cultural understandings, for land-management initiatives, and so on, there is a good chance that the state would have legitimate reasons to continue to provide these subsidies. (More on these questions in Chapter 7.)

A particularly radical possibility is that zoopolis *requires* cellular agriculturists to source cells from farms, in the interests of keeping the farms afloat, ensuring a diversified food system, or similar. Perhaps this is something that the animals themselves would demand via their representatives—and which states must thus consider. I neither endorse this nor rule it out.

This is a pipe dream

Critics might contend that my proposals have little to do with how cellular agriculture is developing. Perhaps it is all well and good proposing how cultivated meat *could* work, but this is different to the question of whether we should support the fledgling cultivated meat industry.

True. This is a work of ideal theory. It is possible that the food systems of today (or the near future) do not (will not) resemble those that I am proposing. Nonetheless, ideal theory has its place (Cooke 2017b). Ideal theory is about exploring possibilities to expand our moral horizons, about giving us a direction to head, and about giving us a yardstick to measure contemporary institutions. These things matter, even if ideal theory cannot—without robust non-ideal theorizing—tell us what to do now.

However, there are reasons to think that the vision I am proposing is achievable. Consider just three factors that lend themselves to the model I have forwarded over, for example, the dystopian visions with which I began this chapter.

First, there are real-world steps towards something like workers' rights for animals (see, e.g., Kymlicka 2017). Second, some small-scale farmers *are* working with the cellular agricultural industry, exploring ways that they can contribute to, or adapt in response to, the new technologies.[16] Third, there

[16] One of the leading figures in cellular agriculture in the UK is Illtud Dunsford, a cofounder of Cultivate UK (an academic society) and Cellular Agriculture Ltd (the first cellular agriculture company in the UK). He is also an animal farmer and the owner of Charcutier Ltd, a slaughter-based meat company.

is the commitment to open access among many of the key players in cellular agriculture.[17]

Many (not all!) of cellular agriculture's champions hope to do good—feed the hungry, help animals, or protect the global environment. Cellular agriculture can do good things, but not, perhaps, if elites control the technology. These questions, and the promise offered for their successful resolution, are another central concern for food tech justice (Broad 2019; Chiles et al. 2021).

It is worth adding that the model I am proposing is at least somewhat *more* realistic than PIB or MOC, given that PIB relies upon non-existent meat-making machines, while MOC relies upon non-existent immortal cell-lines. The animal worker model, while relying on social, economic, and political breakthroughs (as does any envisioned future for cultivated meat), does not rely on any technical breakthroughs other than those needed to scale cultivated meat at all.

Thus, I propose that there are seeds of the decentralized, farmer-friendly vision of cultivated meat production in the contemporary cellular-agriculture industry, and I propose that the idea of workers' rights for animals is no mere pipe dream, but something building upon contemporary politico-legal developments. Although this remains a work of ideal theory—and although I defend the value of ideal theory—I contend that readers should not consider these ideas mere blue-sky thinking.

Concluding remarks

In the zoopolis, we could create cultivated meat using the cells of animal companions or using immortal cell lines, but there is also reason to support farms containing animal workers, protected by workers' rights, as a source of cells. Workers' rights for animals offers the potential for a rights-respecting cultivated meat industry, but, importantly, one that provides a place for liberated animals and preserves routes to sincerely held visions of the good life.

There is a need to fill out details about these workers' rights. While communities must work out many questions in practice, I develop a clearer idea of how we could conceive of animals living on farms as workers in Chapter 6, in which I explore an example not (necessarily) directly related to cultivated meat: Chickens kept for eggs.

---

[17] New Harvest, for example, is a keen supporter of sharing research (Wurgaft 2020, 50).

6
Eggs

We have seen that, in the zoopolis, farmers could keep animals for cells—and that this could be a good thing for consumers, farmers, and even animals. But if we could have rights-respecting cell farms, could we have other rights-respecting pastoral farms? In Chapter 5, I rejected farming animals for corpses. But I have been quiet on less baroque possibilities, including[1] farming animals for eggs and milk. These warrant close examination for at least three reasons.

First, we cannot assume that Chapter 5's considerations suffice. Biopsies of 'beef' cows and milking of 'dairy' cows both raise ethical questions—but they raise *different* ethical questions.

Second, different *foodstuffs* raise different ethical issues. Eggs are not milk, milk is not meat. Perhaps we should be concerned about some-but-not-all these foodstuffs. This has been a recurring issue throughout the book, and needs to be addressed even if my visions of animal 'farming' are rejected. Support for plant-based or cultivated milk need not entail support for plant-based or cultivated meat—and vice versa.

Third, exploring alternative visions of familiar forms of pastoral farming allows us to fill in some of the details of Chapter 5's animal worker model—it allows us, for example, to say more about what it might mean for animals to have a *wage*.

This is a chapter about eggs. For the sake of familiarity, I focus on chickens. Perhaps other birds raise issues that chickens do not, so we should not simply *assume* that the 'farming' of quail, ducks, geese (etc.) for eggs would (could, should) work in the same way. Collecting eggs from non-domesticated birds, meanwhile, raises separate issues. But *much* of what I say here will be relevant to non-chicken birds kept for their eggs. Future work, no doubt, can explore particularities of the 'farming' of other birds for their eggs or the 'harvest' of eggs from wild birds.

---

[1] Could we farm sheep for wool, turning this into the food additives lanolin (E913) or cholecalciferol (vitamin D3)? Or—to step away from food for a second—clothes, bedding, carpets, and the rest? I think so, but I will say no more about it, as wool is not primarily a food product.

I explore eggs for the purposes of illustration. Naturally enough, milk (from cows, goats, sheep …) raises different issues. These are undoubtedly worth exploring, but I leave it to another time to explore dairy farming at length.[2] I hope that my considerations in this chapter will indicate that—*in principle*—dairy farming will be permissible in the zoopolis. But I allow that a zoopolis could coherently permit/support egg farming but not dairy farming—or vice versa.

The chapter advances as follows. First, I set out a preliminary defence of the idea that, in the zoopolis, individuals might permissibly keep 'backyard chickens'. I then explore a series of objections speaking to the specific wrongs that we might think are involved in the collection/consumption of eggs. Having established the ethical permissibility of keeping backyard chickens and consuming eggs, I move on to explore the more institutional question—could we have a rights-respecting egg *farm*? I sketch an image of such a farm, before addressing some objections. I conclude that, not only could individuals acquire eggs from their own chickens, but they could purchase eggs produced by chicken workers.

Such eggs would not be cheap. This does not mean that egg-based foods would not be readily available in the zoopolis, of course—it instead means that eggs would mostly be plant-based or cultivated. But that does not mean that egg farming could not be important, whether for consumers, for farmers, or even for chickens.

Backyard chickens

As we have seen, one of the key differences between old approaches to animal rights and new approaches is that the envisioned future of the new approaches contains animals and humans living side-by-side, while the envisioned future of the old approaches does not. This includes animals living side-by-side with us in our homes.

Although many abolitionists have love and affection for (particular) domesticated animals, they aim for a world in which there are no domesticated animals, as they hold that keeping domesticated animals is *inherently* rights-violating. By contrast, advocates of new, political approaches to animal rights allow that, even in a perfectly just, animal-rights-respecting society,

---

[2] I doubt that farming like this will produce huge volumes of food in the zoopolis, at least compared to plant-based meat (milk, eggs …), cellular agriculture, and invertebrate farming. It would be disproportionate to have multiple chapters addressing different aspects of it. A whole book could be written reimagining a range of forms of farming with animal rights in mind. This is not (quite) that book.

humans and animals will continue to live together. In principle, we may continue to have dogs lazing in our kitchens, cats winding about our feet, and—crucially—chickens scratching in our gardens.

Advocates of new animal rights will ask questions about these animals' treatment. We cannot have a dog lazing in our kitchen if she is locked inside all day, or fed only scraps, or denied medical treatment. And they will ask about these animals' acquisition. We cannot have a cat winding about our feet if she was stolen (catnapped?) from a loving home, or purchased from an unscrupulous breeder. Indeed, animals as fungible property likely sounds incompatible with animals as full and equal members of our society.

But if chickens have come to live in our garden in a respectful way, and the garden is a place of safety and happiness—as a large garden, supported by a conscientious guardian, could be—then advocates of new animal rights need not see their presence as problematic. Chickens, like dogs and cats, can be full and valued members of the family and community. And their presence can benefit other members of the family or community. They can offer companionship and comfort; insight and inspiration; a reason to get up in the morning. But the benefits they provide can be material: they eat pesky bugs; their faeces can enrich compost; their eggs are a source of food.

Given that these chickens will be laying eggs either way, what harm would there be—so might say their guardian—in the occasional omelette? Indeed, they might press, what is the alternative? Assuming (for now) that only hens are present, there will be no chicks. Leaving the eggs will see them decompose, or feed some other animal,[3] which could put the hens at some risk—decomposing eggs may be unhealthy, and egg-eating animals (rats, mustelids, snakes ...) may threaten hens. Binning eggs seems wasteful. Feeding eggs *back* to the hens provides them some benefit[4] (protein, calcium ...), but we could provide these benefits in other ways—other ways, crucially, that would not mean the guardian having to forgo food that (we can stipulate) she finds nutritious, delicious, personally (culturally, spiritually ...) important, or simply convenient.

Against backyard chickens

Granted, a critic might say, someone could live with chickens in the zoopolis and derive *some* benefits from that co-living, such as the benefits of companionship. But, goes this challenge, they could not derive simply *any* benefit

---

[3] These are not so different. Decomposition is a kind of consumption.
[4] This sometimes elicits horror, but it is a common practice among people who care for chickens.

from this co-living. For example, they could not derive the benefit of roast chicken on a Sunday, as killing and roasting chickens wrongs chickens. The benefit of eggs could be—might press the critic—like the illegitimate benefit of roast chicken, not the legitimate benefit of companionship. Taking chickens' eggs might be rights-violating, even while keeping chickens is not. There are a range of ways that our imagined critic might offer this argument.[5]

First, perhaps hens have a right to raise a family. Taking eggs scuppers their chance of doing so, so is illegitimate. Donaldson and Kymlicka, for example, support the idea of offering chickens a chance to raise a family, but this is compatible with taking *excess* eggs from chickens (2011, 138). And, what is more, Donaldson and Kymlicka rightly oppose permitting chickens *unrestricted* opportunity to breed and raise a family (2011, 146), as domesticated animals may lack the ability to self-regulate their reproduction. Chickens living in a backyard could quickly overpopulate, thus 'creat[ing] unsustainable burdens on the scheme of cooperation' (2011, 147). Instead, Donaldson and Kymlicka argue, we should aim for 'socially sustainable' populations (2011, 147). In a household environment, this will often mean denying animals the chance to breed.[6] Even if chickens do have an interest in raising a family, then, this will frequently fail to translate into a right to raise a family.[7]

Second, someone could insist that eggs belong to chickens, not us, and we violate the chickens' property rights when we take them.[8] We could counter that animals cannot own things, or do not own eggs. But let us assume that chickens' eggs do belong to them. We could then observe that chickens (generally) apparently do not mind our taking their eggs, and we might construe this as tacit consent for our taking them. Or we could see chickens as (frequently) willingly abandoning (and, thus, relinquishing a property claim over) eggs.[9] But let us assume that not only do chickens' eggs belong to them, but they would prefer us not to take them. Does that mean that it is illegitimate for us to take them?

Not necessarily. We need not frame our taking of eggs as theft, just as we need not frame the chickens living on our land as squatters. In choosing to live with chickens, we have entered a relationship with them, and benefits of relationships can go both ways—just as the benefits of companionship can

---

[5] The following paragraphs borrow from Fischer and Milburn 2019, 115–117.
[6] Perhaps even if there *is* room for more chickens, it would be appropriate to deny chickens a chance to breed—if, say, there were other chickens in need of a loving home, it would make sense to welcome them rather than bringing new chickens into existence. Nothing depends on this claim.
[7] In the jargon of interest-based rights theory, chickens have a *prima facie* right to raise a family, but often no concrete right because of the pressure such a right would place on the scheme of cooperation.
[8] I conceptualize eggs as birds' property in Milburn 2017a, 644.
[9] I say more about broody chickens later.

accrue to both dogs and humans. We might reasonably think of the eggs as a kind of contribution to the household. Taking this contribution is consistent with affirming the highest moral standing to these chickens, just as expecting human household members to contribute *as far as they are able*— even if that requires a sacrifice of their labour (e.g. chores) or property (e.g. rent)—is consistent with respecting them as full and equal rights-possessing beings. I stress *as far as they are able* because co-living does not *entitle* us to contributions. Some family members cannot contribute (in a given way)— some human family members do not have an income, some chicken family members do not lay eggs.

Let us again note that on this vision of co-living, the chickens are members of our family and community. If we view them differently, the chicken/human relationship starts to sound more like the relationship on the farm discussed below, and thus we must think about the rights of the chickens differently. More on this later in the chapter.

Third, one could again appeal to claims about there being something morally wrong or rights-violating in the consumption of animal products per se. Many such arguments were explored earlier. For instance, in Chapter 3 we saw that 'mixed-messages' arguments are unconvincing in the zoopolis, and in Chapter 4 we saw that conceiving an individual as a potential source of food does not necessarily denigrate that individual.

It is true that we do not see *humans* as a potential source of eggs (*qua* foodstuff), simply because hens' eggs do not have a close human analogue. That said, they have a *fairly* close analogue. Many nonhuman mammals (including, incidentally, herbivorous mammals[10]) practise placentophagy, consuming the placenta after the birth of young. Some humans do, too. This has been a long-lasting, cross-cultural practice, although it can be taboo.

Like all taboo foods, placenta can be valuable (*qua* food) for some people, and the *mere fact it is taboo* may accentuate some of its values. When the chef Hugh Fearnley-Whittingstall created a placenta pate on the British television programme *TV Dinners* in 1998, one diner enjoyed it so much that he had seventeen helpings—although others were less enamoured. Unsurprisingly, the television programme attracted controversy, but Channel 4 defended it by saying it sought to challenge conventional wisdom about food.

As we have seen, the value of food stretches far beyond pleasures of the palate or the chance to indulge in taboos.[11] Some people eat placenta in search

---

[10] Elsewhere (Milburn 2022b, chapter 2), I have argued that considering animals as eaters can offer a valuable lens through which to consider questions about the ethics of human diet.

[11] And perhaps it did with Fearnley-Whittingstall's diner, too; the placenta was his wife's (or, depending on how you think about it, his daughter's). Perhaps he would not have enjoyed the pate so much without this connection.

of health benefits, such as to combat postpartum depression (Benyshek et al. 2018). (The accuracy of these beliefs is a different matter.) It also has spiritual significance: a Gullah belief is that a child born with the placenta covering their face can drink a tea made with placenta to stop them from seeing spirits that would otherwise haunt them (Young and Benyshek 2010, 472).[12] Traditional medicines, too, use placentas. We consequently have every reason to believe that placenta may find a (small) place in the food system of the zoopolis. Recalling our earlier discussion of cannibalism, perhaps one 'cut' of human flesh for which there would be some demand would be placental flesh—in time, perhaps this is something that cellular agriculturalists could produce.

But let us put aside these general arguments about the consumption of animal products—and these further forays into cannibalism—to focus a little more closely on eggs. Eggs differ from milk (although not from meat) in that they exist primarily as something *other* than food. They exist as part of a chicken's reproductive process. There thus might seem to be a kind of ontological mistake in framing them as food. This would match with plenty of popular vegan rhetoric—we need to get away from conversations about 'backyard' chickens, say many vegans, and accept that eggs *are not for eating*.

However, this framing of the 'eggs are not food' argument fails to separate eggs from any number of *plant* (and other non-animal) foods. Nuts, seeds, beans, mushrooms, leaves, roots, grains—none of these things exist first as food. They are grown by plants (or non-plants) for survival or reproduction. It thus starts to sound like this argument misunderstands what food *is—of course* many foods exist first as something other than food. More argument is needed, then, to move from this putative ontological mistake to the putative wrongness of consuming eggs.

But there is another distinctive feature of eggs. Eggs are a form of what Adams calls 'feminized protein' (2017). Protein exists in plants, prior to being cycled through animals' bodies. The production of eggs involves a processing of protein through *female* bodies. Thus, egg production involves a particular *sexualized* injustice: the objectification of female reproductive systems. Thus, the critic might say, the consumption of eggs is inherently illegitimate.

I note that there is a logical leap, here. Let us assume that it is wrong to objectify rights-bearing animals' reproductive systems.[13] It is perhaps right that the objectification of chickens' reproductive systems has the potential to *compound* the injustice in the industrial egg industry and other exploitative

---

[12] The Gullah are an African American people of the south-east United States.
[13] Might the use of surrogate mothers or sperm donors involve *permissible* objectification of reproductive systems? Might (some?) consensual sexual activity involve permissible objectification? These are complex questions.

co-relations between humans and chickens. But it is not obvious that *all* consumption of eggs involves this wrongful objectification. In the case of backyard chickens as imagined, the eggs are *part* of the human/chicken relationship—they are not the sole reason for which the householder lives with chickens.

If they *were* the sole reason for which the householder lives with chickens, that might be objectionable—just as it would be objectionable to have a child solely because they could do chores. It might not be so objectionable, on the other hand, to consider that the children would likely do some chores once able when considering whether to have a child—although potential parents' fixation on chores may seem eccentric.[14]

In the zoopolis, chickens are more than egg-laying machines—they are more than an objectified reproductive system. Equally, they are more than fertilizer-producing machines—they are more than an objectified digestive system. Again, the consumption of human placentas is an apt comparison. The fact that a placenta is eaten is not proof that a woman's reproductive system has been objectified. It *could* have been objectified. Equally, a woman (or female animal) giving birth to a child *could* have had her reproductive system objectified. But not every birth (I contend) involves the objectification of a female reproductive system.

Maybe this is too quick. Adams could argue that, when consumed, the eggs *themselves* are objectified. This is true. But that is no different to the way we objectify *apples* when consuming them. Just as we do not owe anything to apples, so—more on this later—we do not owe anything to eggs. Consequently, the fact that consuming eggs objectifies them is unproblematic.

But there is a difference between eggs and apples in that eggs are the product of rights-bearing beings' reproductive systems, while apples are not. To objectify eggs, then, could be to disrespect the (bearers of the) reproductive systems. But this is a curious claim. We would not find a parallel claim about apples convincing—to objectify apples is not to disrespect the people who grew and picked the apples. Granted, orchardists do not create apples in the way chickens create eggs—but they may expend much physical, emotional, or spiritual labour to do so, just as may chickens. Indeed, (some) orchardists might expend *more* than (some) chickens. It is thus hard to see why we would posit a necessary disrespect of chickens in egg consumption but not of orchardists in apple consumption. If I am right, we should reject the idea that egg consumption is inherently disrespectful.

---

[14] Perhaps not so in communities in which children's labour contributed significantly to household finances.

But perhaps this apple/egg comparison misses the significance of eggs being the product of rights-bearing beings' reproductive systems—something apples are not. But observe that an argument parallel to our egg argument applied to *living beings* sounds tortured. If I objectify a human being, I have (plausibly) wronged that human being. But it would be baroque to suggest that I have therefore wronged the human being's (biological) mother because I have disrespected her reproductive system (and, therefore, her). Similarly, it is a stretch to say that consuming an egg *necessarily* involves disrespect for a chicken—even if, in many real-world circumstances, it does.

Egg farming

In the previous section, I argued that backyard chickens could be a legitimate source of eggs. In Chapter 5, I argued that a model of animals as workers, entitled to workers' rights, could justify pastoral 'farms' in the zoopolis. The workers' rights hypothesized, recall, were as follows:

1. The right to representation by a labour union, including the right to participation (via the union) in workplace decision-making.
2. The right to a decent standard of remuneration.
3. The right to healthy, hygienic, and safe working conditions.
4. The right to time off work including limits to working days and time off work to engage in rest, leisure, and play.
5. The right to retirement and aid in case of unemployment.
6. The right to a choice in one's work.

But let us also recall that these are only the rights that 'farm' animals have *qua* worker. They also have rights *qua* sentient animal, and—one of the insights of the new approaches to animal rights—*qua* community member. The precise details of these rights *qua* community members vary depending on the particular theory of animal rights endorsed. But they, too, will be important in going forward.

Could we scale up our model of backyard chickens to a model of a just egg *farm*? Here, I sketch what such a farm might look like. In so doing, I challenge the implicit presumption of some advocates of zoopolitics that while keeping chickens for eggs is legitimate, the commercialization of egg farming is not (Coulter 2020, fn.4; Donaldson and Kymlicka 2011, 138–139). Of course, without commercialization, it is hard to imagine how such operations could be part of a food system, rather than simply a hobby. Without commercialization, it is hard to imagine eggs on the shelves of supermarkets.

The reason that workers' rights are necessary for egg farms (but not backyards) echo the reason they are necessary in the animal-worker model, but not the pig-in-the-backyard model, of cultivated meat production. The profit motive of the farmer replaces the care motive of the householder, meaning the animals need greater protection. For the farmer, the eggs are the end—for the householder, the chickens are the end. The institution of workers' rights ensures that the chickens are treated as worthy even if they are used as a route to the eggs. To evoke Kantian thought, workers' rights allow farmers to use chickens as a means to an end by ensuring that they are not used as *merely* a means to an end.

Egg farming: A sketch

For ease of exposition, let us call our farm FairEggs, and start with physical appearances. FairEggs would be situated in a large rural space. It would feature laying boxes—for convenience's sake, if nothing else—and roosting boxes, but it would also house an array of natural features and artificial objects for chickens to explore, play with, hide under, and so on. It would have fresh air, clean water, natural ground, seeds, dust, plants—in short, all the things chickens want, that are important for their health, and that allow them to do the things they are drawn to do. It would be large enough that the chickens could form their own friendships and flocks without tripping over each other (literally or figuratively), and, in those flocks, the chickens could engage in the social activities important to them—establishing pecking orders, communicating, negotiating the world.

Farmers could not kill the chickens of FairEggs, excepting genuine euthanasia. At FairEggs—or perhaps a separate facility with cockerels—no chicks would be killed, as machines would be used to identify the sex of embryos after only a few days of incubation (see Galli et al. 2018). Fertilized eggs containing unwanted male embryos would be destroyed. (I address objections shortly.) Less 'productive' hens would live out their time on the farm in peace. 'Retirement' woul not see hens moved—indeed, it would amount to few changes, as their life on the farm is theirs to live; hens will already have their own spaces, flock, and routines. Although I reject this later, we may think that, compared to other working animals, the right to retirement is unimportant to the hens of FairEggs, as their work is integral to their day-to-day lives.[15]

What care would the chickens receive? It is important for farmers to provide for the birds in three key ways. First, farmers must feed the hens. The chickens of FairEggs would have ample, healthful, pleasant, species-appropriate food. Second, farmers would need to ensure that the hens' environment is appropriate for them in the ways described.

Third, farmers must ensure the hens have access to medical care. Farmers could not ignore injuries or illnesses, nor 'cull' 'infected' flocks (although perhaps quarantine would be possible). The killing of ill chickens on FairEggs would be a last resort, done only when in the chickens' own interests.[16] Until that point, those who need it would have veterinary treatment, even if that meant chickens becoming unproductive. (See below.) This would, obviously, be expensive—although perhaps not as expensive as it could be, as many chicken afflictions are caused or exacerbated by contemporary farming practices.[17] Donaldson and Kymlicka write that '[h]ealth care is a right of membership in contemporary societies, and domesticated animals have the right to be treated as members' (2011, 142). This could mean that the state is obliged to provide medical care to hens, but it need not: the cost of the healthcare could be borne by consumers through requiring farmers to pay for insurance.

Ensuring that these levels of support are provided to a decent minimum would safeguard the chickens' labour rights concerning healthy and safe working conditions and rest and leisure time are met.

What of the right to remuneration? Cochrane proposes that 'in kind payments' (that is, the provision of goods or services rather than money) are appropriate for animals (2016, 31–32). Animals need food, healthcare, and appropriate environments. But if the farmer provides these, then perhaps no further payment is needed? I reject this proposal, for two reasons.

First, these chickens are workers, entitling them to more than the minimum afforded to sentient animals. This is a cornerstone of the liberal idea of the worker. While it *may* be the case that some forms of work do not warrant financial reward, these chickens are working for the profit of others. If *any* work entitles one to pay 'above and beyond' what is owed to non-workers,

---

[15] Compare, say, courthouse therapy dogs, who might have separate work and home lives.
[16] And it would not be the decision of the farmer. As explored shortly, chickens' unions could make decisions on chickens' behalf.
[17] E.g., hens are less likely to injure each other if not crowded into cages and less likely to catch infections if not packed into barns.

it is this.[18] But, it might be said, chickens' work does not ask much of them. But this is not that different from many *human* workers. People who love their job might well continue without pay. This does not mean they are not *entitled* to pay, *entitled* to a share of the profits that others make from their endeavours.

It is possible that payment 'above and beyond' could be provided in kind. Farmers could pay chickens by providing them with *more* than the bare minimum they are entitled to—more than the bare minimum of space, food, and so on. This would not be unreasonable were farmers not bound by a profit-motive—a not-for-profit version of FairEggs could perhaps have in-kind payments (cf. Wayne 2013).

This leads to my second reason that in-kind payments are insufficient: The profit-motive of farmers conflicts with the best interests of the chickens, meaning that farmers will be motivated to spend chickens' wages (putatively on the chickens' behalf) in a way that is not in the interests of the chickens. Chickens need *someone* to decide how to spend their wages, but this should be someone motivated to do right by the chickens, not someone motivated by profit.

Here, then, is my proposal: On FairEggs, chickens would be paid a wage on top of the things provided to them by the farmer. Some would go to the chickens' union, as dues. The remainder, however, would be the chickens' 'spending' money. The union could manage this money, paying for things from which all or some[19] of the chickens would benefit, such as features on the farm beyond those provided by the farmer. There are three things to say about this.

First, the human owners of the land would not have complete control over it. The space would not be merely a private workplace, but also the home of many rights-bearing chickens. In opening FairEggs, the farmers would be relinquishing some of their control of the space, just as someone taking on a live-in human member of staff would lose a degree of control over some of their house.[20]

Second, what was (not) the duty of the farmer to provide would have to be worked out in practice, in conversation between the chickens' representatives, the farmers, and legislation.

---

[18] Note that I am making no commitment as to *how much* pay chickens are owed.
[19] Chickens are individuals, with their own preferences and desires.
[20] There are similarities, here, with John Hadley's proposal of overlapping animal and human property rights (2015). On his picture, humans wanting to make developments on wild animals' land need to negotiate with animals' representatives.

Third, the chickens themselves could have their say. If chickens have stopped visiting a patch, perhaps it should be adjusted. If chickens visit a different patch, perhaps it should be expanded. If chickens like a bush, perhaps more should be planted. The space is partly theirs; we spend our wages to better our homes, and so should chickens.

While it can be difficult to interpret what chickens want, the difficulty should not be exaggerated.[21] If unions have *any* role, it is interpreting the needs and wants of their members, and that is as true of chicken unions as human unions. This means chicken representatives would need access to the farm and chickens—again, in establishing a farm, the human owners relinquish some control over the space. Precisely *how much* control is not something that can be decreed from the armchair.

Let us, then, say something about the union itself. This requires a nod towards a wider political structure, although getting too far into details is beyond the scope of this book. True, this book assumes a liberal state that legislates for animals. But this is compatible with a range of possible political structures—different democratic and deliberative mechanisms; different legal and constitutional foundations; different levels of welfare provisions; and so on. As such, any nod towards the wider political structure here should be taken only as indicative. FairEggs, and chicken unions, will 'slot into' different models of different zoopolitical futures in different ways. (Compare Chapter 7.)

For the union to represent the interests of chickens, it needs to be independent from farmers' unions, and needs to not be dependent on the egg industry for money. This would prevent the union from having a perverse incentive to call for increased production against chickens' interests. One solution to this would be to have state support for animal-worker unions in the case of industry failure. This is not unrealistic, as the state will have an income from a tax upon the wages of animal workers,[22] and will have a mandate to act in their interests. It would make sense, then, that they would have a particular duty to support the rights of animal workers. The union, as well as negotiating for change and development, would be able to request independent investigation and litigation. At the extreme, they could call a 'strike', holding back

---

[21] Donaldson and Kymlicka (2011, 103–112) develop the idea of 'dependent agency' found in disability studies to apply to animals, and discuss this in practice at institutions not unlike FairEggs (2015; see also Blattner, Donaldson, and Wilcox 2020). For more on human/animal political communication, see Meijer 2019.

[22] The animal worker literature says little about this, perhaps because wages for animal workers are themselves only speculative. But on the related idea of taxing animals on the profits they derive from their property, see Bradshaw 2020, 128.

(on behalf of the chickens) the products of the chickens' labour. During such a strike, farmers would no longer take eggs.[23]

Chickens cannot vote for their union representatives, so some other system would have to be put in place. To identify this, let us take a step back to the level of *political* representation. The political rights of chickens would need to be realized through political representation—especially if, as I have suggested, chickens' wages would be taxed. ('No taxation without representation.') Naturally, this complex issue cannot be resolved here.

One proposal comes from Cochrane (2018, chapter 3). Deliberative assemblies, made up of humans, could select other humans to represent animals. Cochrane suggests a doubling of political representatives from current numbers to incorporate animal representatives (2018, 51). Were this realized, individuals could be appointed as, say, government ministers for animal workers (or particular kinds of workers). We could even imagine (cf. Cochrane 2018, 48) new 'animal labour parties' formed to focus on the interests of animal labourers (wages, union power, protection/extension of rights, etc.) and contest animal 'elections' on these grounds.

Similar deliberative assemblies could appoint union leaders—invariably non-permanent appointments, or else the chickens face the risk of permanent representation by a poor union they cannot organize against. Alternatively (or additionally), appointments could be overseen by an 'animal civil service', in turn overseen by animal political representatives, and thus democratically accountable. This would allow the animals' representation to be in the spirit of democratic workers' control, even though chickens cannot represent themselves.

Cochrane's 'sentientist democracy' is thus one system that could be developed to provide the kind of representation and support that these chicken workers would require. Perhaps developing other systems would result in something a little different.

Let us return to FairEggs. What of animal workers' right to a *choice* in their employment? We might be concerned that FairEggs has offered the chickens no choice over their employment, and thus fails to respect their rights *qua* workers. If chickens are labourers, we might think, they are *forced* labourers, and this hardly sounds consistent with respect for their rights.

This concern, however, should not worry us. The labour of the chickens here is what they would be doing anyway; we are not asking them to adopt a

---

[23] We could (but need not) frame this as the chickens exercising their property rights over their eggs, something that, I argued earlier, we could think of them generally rescinding. Union workers could collect eggs on strike days, smashing them on the floor for the hens to eat. They should not enter the human food system—consumers, too, should be aware of the strike, incentivizing them to support resolution.

new lifestyle (as we would, say, a guard dog) given that their bodies produce eggs either way. (More on the inevitability of egg production later.)

One obvious rebuttal is that the chickens would not be laying eggs without the farm, as they would not exist. The first thing to say is that they would continue laying eggs were the farm disbanded, or were they taken from it. Consequently, there is a sense in which they *would* be laying eggs absent the farm. The second thing to say is that they did not choose to exist rather than not exist. (No one did.) Perhaps this means that they did not choose this work. But it is hard to see how the case of the chicken is distinct from any other kind of work. *I* did not choose to be born; to need food, water, shelter; to be thrown into a location (temporal, geographic) in which these things are best acquired by working. Granted, if *no one* freely chooses work, then chickens don't. But now the objection seems to have lost its force.

Perhaps chickens do not choose their work insofar as they cannot leave the farm? Paternalistic concern can motivate measures to keep the chickens (somewhat) contained. The natural world is not friendly for domesticated chickens. The chickens (via their union) have a degree of control over their environment, and have considerable freedom over their day-to-day activities. FairEggs thus closely approximates the kind of 'intentional community' that Donaldson and Kymlicka laud as an ideal model for sanctuaries (2015, 63–68; see also Blattner, Donaldson, and Wilcox 2020).

There are important differences between the farm and the sanctuary, however. For example, on the farm, the interests of the chickens are interpreted and represented by third parties, not primary care-givers. This is important because the profit motivation of the farmer differs from the benevolent motivation of the sanctuary-managers. Even in the zoopolis, co-living with animal workers will differ from co-living with residents of sanctuaries.

I take this sketch as preliminary. No doubt it is imperfect. It is based on my own reading of the literature on chicken welfare, including from both animal welfare science and the reports of people who live/work with chickens. I accept that it may have gone wrong. For instance, free-range systems, although associated with many welfare gains, are also associated with higher levels of bird deaths (Elson 2015). Would accidental death be a regular occurrence at FairEggs? Would an indoor version of FairEggs be able to overcome this challenge while still allowing the hens access to all the advantages of an outdoor system? These are interesting questions, and I welcome alternative visions for how humans and animals could live together, and how these alternative ways of living could feed into the food system. But, whatever the imperfections of my own proposal, it does, I believe, provide a glimpse of a kind of pastoral farming very far removed from our own that seems consistent with animal rights.

Objection one: What makes something a farm?

There is a sliding scale between backyard chickens and egg farming. At one end is the vision of a near-vegan living with three rescued chickens, who are as much family members as any animal companion or dependent human child. At the other end is FairEggs, where chickens are treated as ends in themselves not because they are respected family members, but because they have the protection of workers' rights.

In the middle is a range of possible dividing lines. Is someone who lives with chickens *solely* because they like eggs really operating a small farm, rather than keeping some backyard chickens? Could a not-for-profit FairEggs dispense with the need for workers' rights? Must someone with many chickens *necessarily* be running a farm, rather than co-living with chickens in a household?

These are difficult questions. There may not be clear answers, just as there is no hard-and-fast line between a carer and a family member, a tenant and a housemate, or a professional gardener and a neighbour who owes a favour. Lawmakers must draw lines, of course. Different socio-political and legal institutions will come to different kinds of answers. It is not my purpose, here, to prescribe those answers. Instead, I show that not only could these institutions—with a little thought—be consistent with animal rights, but to observe that states might have very good reason to offer that thought.

Objection two: Destruction of eggs

The above sketch of FairEggs mentioned the destruction of eggs containing male foetuses. Might this be rights-violating? Eggs—understood here as shells, albumen, and yolk—are not sentient beings, and thus not rights-bearers. Eggs from backyard hens (or on egg farms) are unfertilized, and so do not *contain* beings that could be rights-bearers. But fertilized eggs are trickier.

Chicken foetuses become sentient prior to hatching: their first sense, touch, is present by the sixth day of development (Tong et al. 2013), so perhaps one can safely assume eggs incubated[24] for five or fewer days do not contain

---

[24] The pertinent question is the length of time since the beginning of incubation, *not* the length of time since a hen laid the egg.

sentient foetuses. If farmers destroy eggs early enough, they do not kill any rights-bearing beings. At most, a sentient being has been stopped from coming into existence, and potential sentient beings have no right to come into existence. (It is surely absurd that I, the male author, am violating the rights of the child I *would* create if I were having unprotected sex with a consenting ovulating partner right now, even if I *could* be having unprotected sex with a consenting ovulating partner rather than writing this book.)

Maybe the wrong is not to the *potential* being, but to existing beings. Let me draw a comparison, borrowing an argument from the literature on immigration. *Even if* it is the case that communities can freely choose not to associate with others—me and my friends do not have to let you join our tiddlywinks league, and, so the argument goes, co-nationals do not have to allow Amy to join their community—it might be wrong to exclude Amy from a political community on the grounds that she is Roma. This is not because it wrongs Amy (*qua* hypothesis, the community can exclude Amy) but because it wrongs Beth, who is a current member of the community and Roma.

As Christopher Heath Wellman (an advocate of the kind of freedom-of-association argument just offered) says,

> we have a special duty to respect our fellow citizens as equal partners in the political cooperative. With this in mind, I suggest that a country may not institute an immigration policy which excludes entry to members of a given race because such a policy would wrongly disrespect those citizens in the dispreferred category.
>
> (Wellman 2008, 139)

By analogy, destroying eggs containing potential *male* chicks, but not eggs containing potential *female* chicks, seems disrespectful to existing cocks. (Note that we cannot dismiss this argument on the ground that the cocks do not care; my disrespecting you does not entail that you feel disrespected.)

But it is not clear that this *does* disrespect existing cocks, even while the Romaphobic policy disrespects existing Roma people. In one case, the implication is 'We do not want your sort here'. In the other, the implication is 'Your sort is welcome, but not needed right now'. It is the difference between rejecting an applicant to a football team because she is gay and rejecting an applicant to a football team because she is a goalkeeper, and the team already has two goalkeepers.

This may sound a problematically instrumental way to treat the chicks—they only get to exist if they are of a sort that fits—but recall that they are *potential* chicks. They do not have rights. Once they *do* exist, they are treated as full and equal members of the community. Indeed, given Donaldson and

Kymlicka's already-quoted words about allowing reproduction only when it is consistent with the continued functioning of the society, perhaps we could say that it is *only because* of this instrumentalizing of *potential* chicks that we can treat *actual* chicks with the full respect they deserve.

Could there be more cocks were the community ordered differently? Certainly. Imagine if chickens were distributed to different households, rather than being concentrated in farms and the backyards of a small number. There could also—if we ordered our society differently—be more hens; more human babies; more hamsters; more of just about anything. Equally, the football team mentioned above could change to a handball team, meaning that all these 'goalkeepers' could find a place. But just as football teams are not obliged to change to handball teams, so we are not obliged to change the scheme of social cooperation to accommodate potential beings. We *may* be obliged to change the scheme of social cooperation to accommodate *actual* beings in need—indeed, the possibility of a zoopolis depends upon this— but we must remember, once again, that we are here talking about *potential* rights-bearers, not *actual* rights-bearers.

Objection three: Broodiness

Much of the argument in favour of FairEggs rests on the assumption that there is no violence in taking eggs from chickens. Chickens usually do not *mind* farmers taking eggs. But some chickens do—specifically, 'broody' chickens. Perhaps taking eggs from broody hens is contrary to their interests, and thus FairEggs has a problem.

This is not the case. Some breeds are less broody than others, and broodiness has been almost entirely bred out of many—in the White Leghorn, for example, it has been reduced 'almost to the vanishing point' (Romanov 2001, 1647). White Leghorns are important for the current egg industry and 'hobby' farmers. Thus, while broodiness might be a problem presently encountered by small-scale farmers utilizing breeds with a higher propensity to broodiness, it is not likely to be seen on FairEggs. Further, chickens are less likely to become broody if the environment does not suit it, such as if eggs are taken promptly (Basheer et al. 2015, 1–2), and thus farmers could take steps to prevent broodiness.

What if broodiness *does* occur? Hobby farmers 'break' broodiness with relatively non-harmful means, like cooling (or repeatedly moving) hens and taking eggs. These may seem unnecessarily invasive if employed to gain a few more eggs. But paternalistic arguments may justify these measures.

Broodiness can be bad for hens: While broody, chickens slow their eating, drinking, preening, and bathing, and pluck feathers from their body.

Paternalistic justifications for 'breaking' would be insufficient for Charlotte Blattner (2020). She argues that animals should have a right to choice over work, which they can communicate in various ways. Perhaps, then, a display of broodiness can be considered a withdrawal of assent of (or consent to)—and a display of dissent towards—the work in question. Crucially, taking animals' choices seriously, for Blattner, means a reluctance to act paternalistically. So, she would say, even if it is bad for chickens to be broody, we cannot use that to disregard their choices.

Presumably, however, we *can* discount their choices if they are bad *for others*. There are at least three senses that broodiness might be harmful for others. First, broody chickens are more aggressive. Second, broody chickens can attract parasites. Third, broodiness itself can be contagious. These could justify breaking broodiness on *non*-paternalistic grounds.

It might seem that this talk about broodiness misses the obvious point: Broody hens display a desire for young, and denying them this opportunity constitutes unjustified interference. As noted earlier, Donaldson and Kymlicka support offering animals, as far as possible, 'autonomous control over their sexual and reproductive lives' (2011, 146). However, this is far from offering reproductive free rein. Instead, they say, we must ask if allowing chickens a given degree of reproductive autonomy is consistent with the continued functioning of the scheme of cooperation.

Is it? Insofar as reproductive freedom would make the community—FairEggs—unsustainable, it seems not. Now, this answer problematically presupposes FairEggs. But such reproduction would also be unsustainable more widely. Our mixed human/animal societies would not be viable if many chickens were having many chicks.

Donaldson and Kymlicka, appropriately, do *not* claim that any given animal has a right to reproduce. Farmers would thus be under no obligation to provide hens with the opportunity to interact with cocks. We need not frame this as an attempt by the farmer to unduly shape the lives of the hens. To echo Donaldson and Kymlicka, due to (perhaps irreversible) human disruption of the reproductive processes of chickens, we have a responsibility to exert some control over their reproduction. This is a greater level of control, perhaps, than we are obliged (or perhaps permitted) to exercise over *other* aspects of their lives.

Thus, and remembering that broodiness would be uncommon, there are three possible responses open to FairEggs. First, farmers could humanely break broodiness for paternalistic reasons. Second—if following

Blattner—farmers could humanely break broodiness for other-directed reasons. Or, third—if following Blattner, *and* the risks of other-harm could be minimized, *and* potential disruption to the scheme of cooperation could be alleviated—farmers could leave the occasional hen to brood, or raise chicks.

Objection four: Selective breeding

To be profitable, FairEggs will employ chickens who have been selectively bred to produce many eggs. These are chickens like those currently exploited for eggs, but—in egg-laying, at least—unlike domestic chickens' wild ancestors. Unsurprisingly, this egg-laying takes its toll. The heavy calcium and energy expenditure that producing eggs requires could be offset (in principle) by providing chickens with enriched foods. But other effects on the hen's body are harder to overcome. Let me focus on the most significant.

Other than humans, domestic chickens are the only animal known to develop ovarian cancer in high numbers. A classic study found some 30–35 per cent of chickens develop cancer by three-and-a-half years of age (Fredrickson 1987), while a more recent study found that over 55 per cent of chickens of one breed had developed cancer by three-and-a-half years of age (Urick, Giles, and Johnson 2009). Significantly, as with humans, the incidence increases with age (Johnson, Stephens, and Giles 2015, 818), and three-and-a-half years is a fraction of a chicken's lifespan. (There are records of chickens living thirty years, but fifteen or twenty years is a more conservative estimate of maximum lifespan.[25]) There is a link between these rates of ovarian cancer and hens' extensive egg-laying. As in humans, 'ovulation rate is a major risk factor related to the risk of ovarian cancer in hens' (Johnson, Stephens, and Giles 2015, 819). Given, first, the harms of ovarian cancer; second, its relation to egg-laying; and, third, the human hand in chickens' genetic history, FairEggs's chickens have been bred so that their bodies do things that are bad for them, leading to high rates of early death and intense suffering.

Perhaps, then, there is something intuitively wrong in bringing these chickens into existence. Indeed, it is common to encounter descriptions of the selective breeding of chickens in activist media. This is painted as part of the constellation of wrongs that animals suffer at human hands. However, it is difficult to see why a given chicken has been wronged by being bred to lay many eggs.

---

[25] https://genomics.senescence.info/species/entry.php? species=Gallus_gallus

To see why, we can look to problems at the intersection of animal- and population ethics. On the one hand, faced with the evidence of ovarian cancer in chickens, we could appeal to a variation of the 'logic of the larder'. This, traditionally, is the argument that vegetarianism is bad for animals, because they would not exist without meat-eating: 'The pig has a stronger interest than any one in the demand for bacon. If all the world were Jewish, there would be no pigs at all' (Leslie Stephen, quoted in Lamey 2019, 177). Reformulated for FairEggs,[26] it would claim that it is better these chickens live a happy life, even though there is a strong chance of ovarian cancer, than not live at all.

Additionally, we might appeal to the non-identity problem. This is the observation that genetic makeup partly makes us who we are. These chickens could not have a better life if they had a different genetic makeup; if they had a different genetic makeup, they would be *different chickens*. It is thus hard to see how the farmer is violating the rights of the chickens (or lowering their welfare). It may be her fault that *these chickens* (rather than *those chickens*) exist, but it is not clear that she has wronged *these chickens* (assuming they would have lives worth living).

There is a literature of responses to the logic of the larder and the non-identity problem. We do not need to get into this, because, as has already been noted, chickens have a right to healthcare. And, because of studies on chickens as an ovarian cancer model, we have documented options for treating them. While this might sound like good news for FairEggs, it is not, as the most straightforward treatment is the use of contraception to stop egg-laying.[27] For example, one study found that the regular injection of a progestin (a hormonal contraceptive) reduced the risk of ovarian cancer by 91 per cent over the sixteen-month experimental period (Treviño et al. 2012). No doubt further studies, especially studies concerned with the hens themselves, could make prevention more complete still.

Aiming to help chickens combat ovarian cancer—and other conditions, such as uterine prolapses—some chicken rescues provide contraceptives. In an article reposted on the website of The Microsanctuary Movement, Ariana Huemer, of Hen Harbour, California, reports that approximately 90 per cent of birds will, if not slaughtered, die from reproductive complications, while 25 per cent of the sanctuary's former battery hens arrive with serious reproductive problems. In addition to surgery, Huemer utilizes contraception to prevent continued laying: 'temporary hormonal implants to shut down their

---

[26] The logic of the larder normally presupposes slaughterhouses. This version does not.
[27] There are non-contraceptive methods, although these are far less effective (Johnson and Giles 2013, 433–434).

ovaries will save their lives'.[28] Similarly, Chicken Run Rescue, Minnesota, treats chickens with deslorilin contraceptive implants, recording the effect this has on the birds.[29]

This means that the choice is not between creating a chicken prone to ovarian cancer or not, but between offering the chicken prone to cancer preventative medicine or not. From an animal-rights perspective, this is not a difficult choice. Chicken community members have a right to healthcare, and this includes a right to inexpensive contraception when it can seriously offset the chance of the development of common, painful, terminal ailments. But this puts FairEggs in jeopardy, because therapeutic contraceptives will result in egg production being reduced to nothing. That is the point. But with no eggs, there is no farm.

What options remain? First, FairEggs could employ chickens who have not been bred to lay body-destroying numbers of eggs. But—especially if the idea is to favour chickens laying a 'natural' number of eggs[30]—this is going to make the eggs of FairEggs an extreme luxury. From laying hundreds of eggs a year, chickens might move to laying, at most, a few dozen. This would surely make eggs so prohibitively expensive that it would not be worth farming them.

Second, FairEggs could make use of some *other* means to limit the risk of hens facing ovarian cancer—and the other ailments associated with egg-laying. But it is not clear what these would be.

A third option seems most viable. Farmers could leave chickens to produce eggs *for a time*, and subsequently offer contraceptives. The ethical viability of this option depends on the health impact. It is not fair to trade off the chickens' health against the farmers' profits; this makes FairEggs sound like a welfarist operation, rather than a serious possibility in the zoopolis. The evidence, however, suggests that this could be viable. The literature on ovarian cancer in chickens reports that the incidence of the disease during the chickens' 'productive' life (i.e. before the chickens are killed in commercial operations) is very low (Johnson, Stephens, and Giles 2015, 819). Meanwhile, the above-noted success of researchers in controlled environments—and of carers on sanctuaries—in treating formerly farmed chickens suggests that beginning to treat chickens with contraceptives at *that* stage of their life can be successful.

---

[28] http://www.microsanctuarymovement.org/hen-harbor-advocates-for-hormonal-implants-to-save-rescued-battery-cage-hens/
[29] http://www.chickenrunrescue.org/No-Such-Thing-as-a-Harmless-Egg
[30] I.e., the number of eggs that the domestic chicken's immediate wild ancestor, the red junglefowl, would 'naturally' lay.

As such, FairEggs's chickens could retire at eighteen months, at which point they would begin hormonal or contraceptive treatment to radically reduce their egg-laying, protecting them from reproductive ailments. (This proposal has a symmetry with current practices, as egg-laying chickens are currently *killed* at around that age; their productivity drops 'naturally'.) This allows a balance between the interests of the chickens in not facing serious reproductive health problems and the interests of the farmers in being able to produce eggs. Farmers would still get the amount of 'laying' they have currently, while the birds would get the healthcare they need.

Concluding remarks

In the zoopolis, there would be at least four possible sources of eggs. One would be beloved companion chickens. These would not be available for sale—if they were, incentives for animals' carers would change, and the chickens would be entitled to workers' rights. The second would be chickens living on farms, who were protected by robust workers' rights. These would not be *cheap* eggs. But, nonetheless, the value of fresh eggs and the institution of the farm might be sufficiently important to many individuals that they would be willing to pay these prices.

Given that egg farming could be just, the zoopolis would have no business outlawing it. The zoopolis may also—given the importance that people may place in, first, access to eggs (see Chapter 1) and, second, the farm (see Chapter 6)—have reason to support it. The fact that some people value access to eggs and the possibility of a certain kind of farming life is the key reason that it is *worth* setting up these complex institutions—workers' rights, unions, etc.—rather than just banning egg-farming altogether. Such a ban would, for no good reason, cut off many citizens' vision of the good life.

Do people value access to eggs this much? In a way, that is not for me to say. But there is every reason to believe they do, as eggs are associated with ways of living that people value. Fresh eggs evoke a quiet, bucolic life (known idiomatically, appropriately enough, as *the good life*[31]), and form an important part of certain professional goals. The master baker, producing artistically or symbolically valuable cakes, might lose her craft without

---

[31] *The Good Life* was a British sitcom about a middle-class couple trying to become self-sufficient—but presumably the association between the phrase *the good life* and a particular kind of rural-inspired life predates the programme.

eggs. And the people buying her products, whether because they are foodies, or because they want to mark their most valuable moments (weddings, birthdays, etc.) with cake, might say that these make all the difference. The interpersonal, cultural, and religious significance of eggs is shown by Easter traditions in (culturally) Christian communities. Eggs are decorated, hidden, rolled, smashed, and—of course—eaten. If it isn't Easter without eggs, the loss of eggs would be tragic for those who place significance in Easter. So, yes—whatever one's own relationship with eggs, there is every reason to believe access to eggs might be significant for many, and so every reason to preserve egg production *if we can do it respectfully*.

Let's not forget, either, that the zoopolis's reason for supporting egg farming can be the animals. The farm gives liberated animals a place, rather than condemning them to (near) extinction. I even think it possible that the farm is a more desirable home than the sanctuary. Working chickens would have money and freedom to—through an intermediary—spend it on their space. Chickens living on a sanctuary, to offer one comparison, might not be afforded such opportunities. And if the work on the farm is *good* work, it may be valuable for its own sake. It may be better, for animals themselves, to have *good* work rather than *no* work (Cochrane 2020).

But what of the (many) people who do not live with chickens, and are not able (or do not want) to spend a great deal on farmed eggs? This brings us to the third and fourth sources of eggs in the zoopolis: Cellular agriculture, and plant-based egg analogues.

Clara Foods—one of the first cellular agriculture companies to have a public face—pursued egg whites using fermentation technology, and many other companies also now work on eggs. Cellular agriculture offers the chance to have products chemically indistinguishable from eggs but which have never been inside the body of a chicken. The first cultivated eggs to market are not likely to look like eggs—just as the first cultivated meat to market was not a sirloin steak. Soon, perhaps, consumers will purchase bottles or jars of egg whites, egg yolks, or mixed whites and yolks.

In time, there is no reason that cultivated eggs could not closely resemble eggs—complete with a plant-based or cultivated shell—if that is what consumers want. Bottled egg is not quite right for making a soft-boiled egg with soldiers, perhaps, even if it is great for baking. A fried egg might be tricky to make with bottled egg, even while an omelette might be easy.

By this point in the book, I have explored the various ethical questions that these plant-based and cellular agricultural products might raise. I explored eggs in this chapter; the ethics of replicating animal products throughout discussions of plant-based and cultivated meat; and the challenges of precision fermentation in my exploration of cellular agriculture. Indeed, at this point in the book, I have explored all the foods that I intend to address at length. The remaining questions, I contend, are broader: they concern how we can bring about, and sustain, food systems. I turn to these questions in Chapter 7.

7
Creating and sustaining just food systems

Let us take a step back. From Chapters 2–6, I paid close attention to the ethical questions raised by assorted foodstuffs and food production methods. I have done this to suggest that these products and methods could (should) find a place in the food system of the zoopolis. But the role of political actors has never been fully in focus. This may seem remiss.

I have talked of the zoopolis (*qua* state) permitting or supporting particular industries as a matter of ideal theory, but said little about what that means. And I have *deferred* mention of what my arguments mean for political agents today. This is a book of ideal theory, and even if—as Rawls (1999, 8) believes—ideal theory has priority over non-ideal theory, it is non-ideal theory that really matters. It is non-ideal theory that allows us to free the downtrodden from the yoke of injustice today,[1] and it is non-ideal theory that lets us work towards a brighter tomorrow.

In this chapter, I address the questions of the creation and maintenance of a just food system. However, I forewarn readers that they may find my exploration unsatisfactory, or at least incomplete. Full answers to these questions are beyond this book's scope.

When it comes to determining the actions of the zoopolis in relation to the food system, we need a fuller theory of justice than I commit to here. I have presupposed a liberal state that respects animals' rights, but this is compatible with many visions of liberalism and of animal rights. I hope, for example, that my conclusions will be amenable to both right-leaning and left-leaning liberals, even though these liberals offer quite different accounts of how, and the extent to which, the state should involve itself in the food system. Thus, my answer to this question will be by way of a partial, illustrative review of tools and approaches in the liberal lexicon. I will not decree precisely how the zoopolis *should* interact with the food system. Instead, I will show that there are many ways it *could*.

---

[1] Yokes, of course, symbolize subjugation by being literal tools of animal oppression.

Food, Justice, and Animals. Josh Milburn, Oxford University Press. © Josh Milburn (2023).
DOI: 10.1093/oso/9780192867469.003.0008

When it comes to determining the conduct of actors—states, activist organizations, individuals—*today*, our considerations move from the ideal to the non-ideal. I would need considerable space to fully develop a non-ideal theory. Instead, I offer a brief and candid (although not, I hope, polemical) reflection on what I *imagine* a non-ideal complement to this book's ideal theorizing would be.

Again, I hope to avoid alienating sympathetic readers who favour alternative activist approaches. Just as the conclusions of this book are compatible with a range of different approaches to liberalism, so they are compatible with a range of different non-ideal approaches. I welcome further work on what my ideal theoretic considerations would (could, should) mean today—even if the conclusions of this work differ from my own.

Permitting and supporting

Throughout this book, I have noted food systems and food production methods that I think the zoopolis should be *permitting* and *supporting*. But I have said little about what these words mean.

Permitting

In the abstract, *permitting* something simply means not forbidding it. In practice, it might mean more. For example, permitting the farming of mussels could oblige the state to fund studies into mussel sentience—let us remember that, in the zoopolis, a permission to farm mussels would be an *exception* to general laws against killing animals. Maybe ongoing studies would make mussel farming permanently provisional—as any potentially harmful industry should be.

Or, take the kind of 'farming' of sentient animals discussed in Chapters 5 and 6. I argued that this farming is consistent with justice *assuming* particular socio-economic structures and legal systems—assuming systems in which animal labour rights are recognized, in which animals have political representation, and so forth. It is incumbent upon the state, insofar as it 'permits' this farming, to establish and maintain the requisite institutions.

Permitting, then, can be more than refusing to prevent. Instead, it may require oversight, or institutional adaptation. I am reluctant, however, to dictate precisely what this oversight and these institutions should look like. The zoopolis will have to work out much of this in practice, and while the input of experts will be crucial, political philosophy is but one relevant area.

Supporting: A puzzle

What does it mean for a liberal state to *support* elements of a food system? My argument is that the zoopolis should support certain non-vegan elements of a food system (or support whole non-vegan food systems), not merely permit them. Or, at least, that the zoopolis has good reasons to support them. Here, we see the diversity of 'liberal' conceptions of the state, and the diversity of argumentative strategies upon which liberals can draw.

Supporting: Smaller states

'Minarchist' liberals (at the 'classical' end of liberalism's spectrum) will say that the food system is not the state's business.[2] The state should leave the food system to the free market. Although minarchist liberals will allow that the state must oversee the food system to prevent rights violations, they will say that the state should not be 'encouraging' any part of the food system.

But this is too quick. Even the most minimal liberalism allows that the state has some business *buying* food, directly or indirectly. Why? The state will be responsible for (some of) the following: prisons, schools, policing, courts, houses of parliament, embassies, institutions for aiding those in dire need … And, at all such institutions, people will be fed. (In addition to employees, there will be prisoners, students, jurors, experts, dignitaries, refugees …) Thus, the state must make decisions about which elements of the food system it supports in those purchase decisions. There is more at stake than mere efficiency. In these decisions, the state (in offering material and symbolic support to certain industries) is saying something—it must decide *what* it says. This decision should be responsive to the considerations canvassed in Chapter 1. Perhaps the state should 'say' it supports certain non-vegan industries.

John Stuart Mill, an important classical liberal, offers support for this vision. He writes that it is 'the duty of the State to consider, in the imposition

---

[2] Even if some commentators associate animal rights with the political left, there is no reason that small-state liberals could not be supporters of animal rights. I have argued (Milburn 2017b) that we can read Robert Nozick as a kind of reluctant animal rights theorist, for example. Nozick is a libertarian, not a liberal, but the libertarian/liberal border blurs.

of taxes, what commodities the consumers can best spare' (Mill 2008, 112)—and thus raise revenue by taxing 'sparable' goods, if such 'indirect' taxation is necessary.

But this works in reverse, too. When the state is deciding which 'commodities' to favour, it should consider which consumers can spare, and *which they cannot*. What foods can people spare? Central to liberalism (including for Mill!) is a recognition that we should listen to people when they tell us what is good for them—what *they* 'can best', and cannot, spare. Perhaps, for many people, the foods that they cannot 'spare' are animal-based foods, meaning that the state should actively support these industries when they use their purchasing power. (Mill's imperative need not be decisive. A duty to *consider* what consumers can spare is not a duty to exclusively favour whatever they cannot spare.)

What is more, the state has a duty to ensure that citizens have access to food. I have stressed that there is a *right* to food, and take it that (most?) liberals will agree. (We can express a 'right to food' in different terms. Rawlsians, say, might talk in terms of a 'social minimum', or goods allowing individuals to fulfil 'basic needs'.) If someone is starving to death, they are almost certainly the victim of an injustice. If the free market ensures that all have access to food, the classical liberal will say, then perhaps non-intervention is appropriate. But if it does not, the state must provide the food themselves, or push the food system in the right direction. Either way, it must ask the same questions. What food should it purchase? What part of the food system should it push? The same answer, I think, applies. It must ask what it should 'say', and must consider what citizens can (and cannot) spare.

In Chapter 1, I argued that the state's reasons for championing non-vegan food systems point towards concerns of *justice*. The preservation, protection, or encouragement of certain non-vegan foodways might be important for food justice concerns. For example, given that meat is a nutrient-dense food, a cultivated meat industry may be valuable for aiding the food insecure, and thus worth encouraging (preserving, establishing) for the sake of providing aid. For geographic reasons, pastoral agriculture (whether with sentient 'worker' animals, or with non-sentient animals) may be accessible to people for whom arable agriculture is not, providing subsistence and livelihoods. This, too, may give the liberal state every reason to materially support such practices.

On the other hand, respect for *animals'* rights may provide the impetus to establish or support alternatives to plant-based food production. If cockling has significantly lower impact on (sentient) wild animals than soybean farming, that gives the liberal state clear reasons—even clear *justice-based* reasons—to support cockles over soybeans. This includes, perhaps among other reasons, in the state's own purchasing decisions.

These decisions depend upon myriad empirical factors and competing values. Decision-makers must explore and weigh these, and I will not try to prejudge said exploration and weighing. My point is that, even for the minarchist, there may be plenty of ways the state could (should) 'support' a non-vegan food system. And if minarchist liberals can get behind supporting a particular food system, *any* liberal can.

Supporting: Larger states

Let us turn to larger visions of the liberal state. Striking for current purposes[3] are what we can call 'perfectionist' liberal approaches. Perfectionist liberals stress that the good life for humans[4] consists in something objective, and they claim that it is the role of the state to help people towards that objective good life. They are thus generally supportive of a comparatively large state. But perfectionist liberals remain *liberals*. They are committed to providing people with space to realize their conceptions of the good. Indeed, they help *create* that space.

Take Martha Nussbaum (2007), who offers a recognizably political (and recognizably liberal) account of animal rights with perfectionist underpinnings. She argues that humans and animals have certain 'capabilities' that they will be free to realize in a just state. Some people will realize their capabilities through animal-based foods. For example, Nussbaum's 'central human capabilities' includes protection for people creating and experiencing 'works and events of one's own choice' (2007, 76): think of the foodie, the pastry chef, or the person celebrating Thanksgiving for whom animal products are central. It also includes a chance of living 'in relation to animals, plants, and the world of nature' (2007, 77): think of the agrarian, the nationalist, the conservationist for whom animal farming is central.

But we do not need to list capabilities to understand the role that the perfectionist liberal state might take in shaping, maintaining, *supporting* a food system. Central to perfectionist *liberalism* is the belief that 'the state [should] legislate in ways that promote its citizens' autonomous flourishing in all aspects of their lives' (Barnhill and Bonotti 2022, 73)—at least in those

---

[3] Not because I am a perfectionist—I am not—but because perfectionists see the state as having a distinctive role in realizing citizens' conceptions of the good.

[4] And animals?

areas 'that structure the contours of' people's lives 'in their most fundamentally defining ways' (Powers, Faden, and Saghai 2012, 9). For the perfectionist liberal, then, (some) freedom becomes an end in itself. And this means actively legislating to protect freedom of choice, when the choices in question are central. Take, for example, choice over work. Protecting this freedom, for the perfectionist liberal, naturally means preventing forced work. But it might also mean—given the centrality of work in people's lives—ensuring that we have some sufficient level of choice over our work, and access to (fair competition for) the kinds of work central to our own visions of the good.

Particular animal products (or particular jobs, subcultures, etc. associated with animal products) are important for people's conceptions of the good life, as argued throughout this book. This is—for the liberal perfectionist—a reason for the state to actively create or maintain a non-vegan food system. Tellingly, for example, Nussbaum (2007, 402) is nervous about a widespread shift to veganism, worrying about how it might interact with the capabilities of humans—even while recognizing that justice demands radical shifts away from current practices of animal farming.[5] Perhaps a food system closer to the one I have explored would resonate with her approach more than a plant-based system.

A zoopolis built on perfectionist liberal principles, then, could support and champion a non-vegan food system in many ways. Valued industries might be supported through subsidies or advertising campaigns; trade deals might support a valued home-grown industry for which there are few local buyers, or a valued food for which there is no local industry; members of disadvantaged communities might be supported in their efforts to find good work with food; and so on.

Social democratic visions of liberalism—which can, but need not, be perfectionist—might support the nationalization and democratization of the food system. An involved liberal state could 'encourage' a certain industry or food system by literally running it. In so doing (says the social democrat) they could provide valued foods, jobs, lifestyles, landscapes, and the rest; realize the demands of food justice; minimize incidental harm to animals; and provide a home for liberated animals. Some commentators, for example, have called for a 'socialized' cellular agriculture industry (e.g., Dutkiewicz 2019). Although I confess to nervousness about *this* level of state involvement in the food system, the vision is compatible with my account.

---

[5] Her focus in the cited passage is on the health impacts of widespread dietary shifts. Her broader point is about tragic conflicts in realizing justice for humans *and* animals.

These suggestions indicate more substantive ways that a liberal state—specifically, a perfectionist liberal state, and/or social democratic liberal state—might support a non-vegan food system.

Supporting: Public reason

Perhaps it is not *just* the perfectionist liberal state that could support industries, practices, products, and more important for people's conceptions of the good. Non-perfectionist liberals support state *neutrality*. That is, they claim that the state has no business declaring *this* or *that* conception of the good correct.[6] Rather than appeal to conceptions of the good, states must justify laws with reference to *public reason*. These are reasons that are, in some important sense, shared by members of the community.

What *specific* (kinds of) reasons are 'public' will vary between conceptions of public reason. In the abstract, however, the idea is compelling. There is a difference between the claim that the state should support the growing of carrots because carrots are delicious, and the claim that the state should support the growing of carrots because without them people will starve. Reasonable people are going to disagree about the delicious-ness of carrots—and even if they do not, they are going to disagree about the importance of deliciousness for public policy. But while reasonable people might disagree about whether carrot-growing is necessary to ward off starvation, they can surely agree that warding off starvation is (would be) a good reason for the state to support carrot-growing.

Implicitly, although I have not couched it in these terms, I have been appealing to public reasons throughout this book. I accept, though, that I have controversially included within the domain of public reason certain somewhat novel justice claims: namely, animal rights. Many reasonable people, advocates of public reason will observe, reject the idea that animals have rights (Zuolo 2020). In response, I echo Cochrane's observation that liberalism needs to adopt a '*new* form of reasonableness': one that sees a rejection of sentient animals' worth as unreasonable, just as mainstream liberal conceptions of reasonableness reject those visions which refuse to affirm human persons' worth (Cochrane 2018, 104–105). This is a big ask. But, again, I am engaging in ideal theory.

In what ways could a liberal account centred on public reason justify state support for non-vegan food systems? The first thing to note is some public

---

[6] The liberal state may—must—restrict people's pursuit of the good when it is contrary to justice, as in Chapter 1's warlord case. But the state need not declare the warlord's understanding objectively wrong.

reason liberals hold that the need to appeal to public reasons in justifications applies only to coercive measures, and/or measures concerned with 'constitutional essentials'. This would mean that policymakers can support *other* kinds of decisions with non-public reasons (Barnhill and Bonotti 2022, 125). So, for example, a liberal state could offer subsidies to cultivated meat without appeals to public reasons—*assuming* this vision of public reason; *assuming* that lots of people really like meat; and *assuming* that subsidies for cellular agriculture are neither coercive nor a matter of constitutional essentials.

Here, however, I follow Anne Barnhill and Matteo Bonotti in affirming a broader account of the role of public reason. (I do this for the sake of illustration—not because of my own views.)

Even on this account of public reason, the state could justify policies benefiting one conception of the good over another. In the jargon, the state must display justificatory neutrality, not consequential neutrality (Kymlicka 1989): it must justify actions without favouring any conceptions of the good life, but it may take actions that benefit some conceptions of the good life over others.

Let us note, then, that there is an array of public reasons that could justify state involvement in the food system, and that these could be used to support (elements of) a non-vegan food system. For Rawlsians, public reasons include 'reasons referring to basic rights and liberties (e.g. freedom of thought and conscience, freedom of religion, freedom of association, etc.) as well as reasons referring to equality of opportunity and the common good' (Barnhill and Bonotti 2022, 68)—plus reasons referring to the necessary conditions for these (Barnhill and Bonotti 2022, 158)—as well as 'the methods and conclusions of science when these are not controversial', 'plain truths now widely accepted, or available, to citizens generally', and 'presently accepted general beliefs and forms of reasoning found in common sense' (Rawls, quoted in Barnhill and Bonotti 2022, 68).

As Barnhill and Bonotti argue—although focused, in their case, on healthy eating measures—this account gives policymakers a range of reasons to support policies directly favouring some foods over others. Let me give three examples.

First, a food system is a necessary condition for the achievement of humans' assorted rights and freedoms. Without food we will die, or at least be very unwell; without life and health, we can have no (say) equality of opportunity (Barnhill and Bonotti 2022, 142). Insofar as a non-vegan food system would be more effective at realizing goals related to food security (and public health), that is a good public reason to support it over a vegan food system. And it might be more effective. Animal products are nutrient-dense, and thus valuable for those who are food insecure; animal products are seemingly more accessible than plant-based products for some with health challenges; and the land (skills, resources) of some food insecure people may not be well-suited to arable agriculture, even while it *might* be suitable for (rights-respecting) forms of animal farming.

Second, the food system—as a series of major interlinking industries that undergird all *other* industries—contributes to communities' financial good, which is a part of the common good. And a healthy, secure food system contributes more substantially to society's financial, and thus (all else equal) common good (Barnhill and Bonotti 2022, 142). We can imagine a variety of circumstances in which a non-vegan food system will be more economically viable than a plant-based food system. Again, those cases in which land is decidedly unsuited to arable agriculture, but *could* be suitable for rights-respecting pastoral agriculture, provide an example. Similarly, policy-makers could offer economic arguments for the value of a food system resting on a multitude of different food sources, rather than a comparatively small number; of animals remaining part of rural economies and cultures; and for the foundation/maintenance of a high-tech, high-skilled cellular agriculture industry.

Third, central to questions about food systems are questions of sustainability (Barnhill and Bonotti 2022, 17–18). There are all kinds of ways in which our global food system impacts the planet, many—although not all—of which are scientifically uncontroversial (in the Rawlsian sense). At the extreme, if our planet becomes uninhabitable, no one will be able to exercise their basic freedoms, and there will be no common good. But there are public reasons to be concerned about the environmental impact of our food system even without appeal to an uninhabitable earth. Pollution, environmental degradation, climate change, biodiversity loss, and similar issues impact our economies, health, food supply, security, and more. If a non-vegan food system could have environmental benefits over a vegan food system, that is a good public reason for the state to offer it support. For example, both bivalve farming and (reimagined) regenerative ruminant agriculture could be consistent with the models of animal farming imagined in this book—and both could have significant environmental benefits.

But we may want to go further than these—relatively conservative—public reasons. After all, and although they have recurred throughout the book, I did not focus on reasons of health, economy, and sustainability in Chapter 1. Why should concerns about the shape of the food system have to be phrased around (fairly specific) concerns about (say) equality of opportunity and the 'common good' of the economy? Can we not just talk about the values that motivate us to challenge/support and change/maintain the food system in the first place?

Present in public reason, according to Rawls, are 'various shared political values' (Barnhill and Bonotti 2022, 141), including, crucially, 'the values of political justice' (Rawls, quoted in Barnhill and Bonotti 2022, 141). (Some public reasons quoted earlier—such as equality of opportunity—are Rawls's *examples* of values of political justice.) But if the demands of justice are public reasons, then the Rawlsian conception of public reason[7] is much wider. For example, policymakers can appeal to food justice concerns *directly*, rather than via (say) the economy. Food security, food sovereignty, good work with food, the preservation of at-threat food cultures, and so on become public reasons. And all of these are reasons that could—depending on myriad particularities—support (elements of) a rights-respecting non-vegan food system.

But political justice makes public another set of reasons. Animals' rights are *themselves* matters of justice. And if animals' rights are a matter of justice, we can appeal to them directly as public reasons. And if *that* is so, then perhaps policymakers can appeal to Chapter 1's animal-based reasons in favour of a non-vegan food system, justifying policies supporting said food system. These were, recall, that arable agriculture harms animals, and so respectful non-vegan sources of food may be more animal-friendly than some forms of arable agriculture; and that respectful non-vegan food production methods can provide a *place* for animals in a way that a fully plant-based food system cannot.

Direct appeals to political justice thus (in principle) open public reason to three of the four 'troubles with veganism' canvassed in Chapter 1. What of the fourth—the fact that many people's conceptions of the good directly or indirectly rest upon non-vegan food systems?

It is tempting to think that we should simply accept that 'x is valuable for some people's vision of the good life' is a public reason. After all, we all have a conception of the good life, and it is central to our lives. And, as a society, we already do a great deal to help people realize their pursuit of the good. This can mean more than simply protecting them from undue interference. Education, healthcare, social security … What are these for if not to help people realize and achieve the good? Indeed, what is the point *of society* if not to help people realize and achieve the good?[8] But states can go further. Perhaps the

---

[7] Rawlsian, if not Rawls's.
[8] We cannot overstate the importance of conceptions of the good in setting up society at all—society being, for Rawls, 'a cooperative venture for mutual advantage' (1999, 4). The contractors in Rawls's original

arts, sports, and—of course—*food* might legitimately receive state support of various kinds, too, if they are important for realizing some people's pursuit of the good life.

But maybe this is too quick. Maybe some reasonable people will deny that it is the role of the state to help people pursue their conceptions of the good—even if those conceptions are widely shared. But what might be *harder* to deny is that certain food-related policies can adversely impact human groups, especially disadvantaged groups, and this kind of impact can be a public reason to oppose them. Certain policies place unreasonable burdens upon people, even if they have come about neutrally, in a non-discriminatory way. When a food policy

> means not being able to pursue one's life plan (or being able to do so only at an excessive or unreasonable cost), and/or when those suffering from the disproportionate burdens resulting from a [food-related policy] are mainly members of an already disadvantaged group (which may also suffer from the disproportionate burdens resulting from other policies, perhaps due to structural injustice), then we are in the presence of strains of commitment that undermine public justification for the policy.
>
> (Barnhill and Bonotti 2022, 173)

Consequently, public reason liberals have every reason to pay close attention to the ways that non-vegan foods and food systems are important for people's conceptions of the good. Non-vegan—but rights-respecting!—food systems may be able to overcome the 'strains of commitment' that realizing animal rights throw up, given animal rights' impact on people's ability to pursue sincerely held conceptions of the good.

I could say more about public reason. Nonetheless, I hope this exploration has indicated directions that public reason liberals could follow to justify support for non-vegan food systems. Drawing from Barnhill and Bonotti (and Rawls), I have argued that:

1) On some visions of public reason, many questions about food systems need not appeal to public reasons, meaning that liberal states could draw upon *many* reasons to pass food-related laws.

---

position—the 'people' who decide what justice consists in—aim to protect the pursuit of the good life. They aim to decide 'which conceptions of justice are most to their advantage' (1999, 123), with 'advantage' decided relative to 'their' conception of the good.

2) Rawls's conception of public reason points to many values (e.g. public health, the economy, sustainability) that could speak in favour of rights-respecting non-vegan food systems.
3) Public reason includes the demands of justice, and thus—I contend— the demands of food justice and animal rights, meaning that arguments resting on these intellectual resources for rights-respecting non-vegan food systems are 'public'.
4) The existence of diverse conceptions of the good speaking in favour of rights-respecting non-vegan food systems may be a public reason to support these systems, but, in any case, the strains of commitment can speak against policies failing to allow people room to pursue their conceptions of the good.

Supporting: Competing visions

It is not my purpose to decide between liberalism's competing conceptions. It is simply to observe that, whatever *liberal* means in a community, there is reason to believe that the state could permit *and* support (elements of) a non-vegan food system. But what *support* looks like, and what justifies that support, may vary.

Does this conclusion extend beyond liberalism? In Chapter 5, I skirted the question of whether it is possible for non-liberal states to fully respect rights. But, assuming that 'animal rights' make sense in non-liberal societies, such states, too, may have reason to champion non-vegan food systems. I welcome further developments of the arguments I have offered in this book utilizing non-liberal intellectual resources.

A socialist zoopolis will presumably favour food production methods that involve lower impact on animals, and that find good, meaningful work (as defined by the state) for human and animal citizens. A cosmozoopolitics— rejecting the idea of a 'state' in favour of a cosmopolitan political order— will still have systems of food governance at local, 'national', and 'international' levels, so will still have to grapple with questions about what food systems legislators should permit and support. In principle, it could draw upon the same considerations as a liberal zoopolis to make these decisions.

The argument might not extend to all possible animal-friendly socio-political structures, however. A libertarian zoopolis presumably would not interfere with the free market in the way that 'supporting' one food system over another would entail. An anarchist zoo'polis' would have no state to do the supporting at all.

What do these arguments mean today?

The forgoing reflections on how the state could (not) support food systems remain firmly in the realm of ideal theory. Readers might fairly contend that the *real* questions about creating and maintaining food systems are not questions of future states, but questions of our actions now.

It is not my primary purpose to offer direct guidance on how we should live today, either as individuals existing in an unjust world, or as communities seeking to shift from unjust to just structures. But these questions are important—indeed, perhaps *more* important than questions of ideal theory (Rawls 1999, 8). Whether the reader is sympathetic to what I have argued or not, they will surely wonder what accepting my arguments mean for us now—*qua* states, *qua* eaters, *qua* political agents.

My reflections on this—to echo my comments from the start of this chapter—will be opinionated and speculative, but hopefully neither polemical nor empty. I hope to start a conversation and share some of my own thoughts, rather than reach decisive answers on what my ideal theorizing means for real-world progress. I begin with what my case means for individual diets, before turning to consider how states in today's world might move towards the ideal identified, and how (animal) activists might appropriately (that is: permissively, effectively …) push the state in that direction.

Individuals

If there are means of producing meat, milk, eggs, and so on that are respectful of animal rights, that would surely mean—someone might say—that it is permissible to eat those foods today. The short answer, as with so many things, is *Yes, but …*

First: *Yes, but* these foods might not be available today.[9] No farmers produce eggs in accordance with the animal worker model sketched in this book, meaning no readers have access to such eggs, even if readers *might* have access to rights-respecting eggs laid by their own chicken companions. Meanwhile, few readers (depending on when and where they are reading) will have access to cultivated meat—and if they do, it may be cultivated meat produced using

---

[9] That is, the foods may be unavailable, or their current production may be rights-violating.

FBS, or in otherwise unjust ways. On the other hand, I suspect most or all readers *will* have access to plant-based meats, meaning that they might fairly conclude that, if this book's arguments are right, they might permissibly eat them today.

But this leads to a second answer. *Yes, but* that conclusion presupposes that there are no *moral* reasons that speak against these products' consumption today. Even actions which respect rights (and are consistent with the demands of justice) might be immoral, or have moral reasons that speak against them. There may even be arguments suggesting that products morally innocuous *in the zoopolis* are morally questionable *today*.

I explored some reasons that may be of this character in Chapters 3 and 4. For example, there may be good moral reasons against us eating any animal products or plant-based meats *today*, on the grounds that doing so sends mixed messages: My eating *this* rights-respecting meat might lead those around me to eat *that* non-rights-respecting meat. This fact might give me a good moral reason to not eat *this* meat, even though I am (*qua* hypothesis) violating no rights in doing so.

Of course, this worry does not apply in the zoopolis, where no one would interpret my eating of rights-respecting meat as endorsing the eating of non-rights-respecting meat. There would not even be non-rights respecting meat available. Thus—assuming this kind of argument holds up—it might be immoral to eat meat (milk, eggs, honey …) in today's world, even if it is produced respectfully, but *not* immoral to do so in the zoopolis.

Thus, someone could hold that the arguments of this book are correct, yet nonetheless argue that *for now* we should be vegan, and/or (for now) eschew plant-based meats (and similar). I remain vegan, despite being convinced of the arguments of this book. (Although I eat plenty of plant-based meats, cheeses, and more.)

That said, I am not convinced these arguments hold up. I am far from committed to the claim that we must be vegan today *even though* ideal food systems would (might) not be vegan.[10] I am simply observing that there is no straight line from 'ideal food systems would be non-vegan' to 'it's permissible to be non-vegan today'. Instead, we need to engage in some careful moral reflection. Thankfully, philosophers have worked on these questions. Indeed, up to now, animal ethicists have written more on the moral questions I am

---

[10] Perhaps I am a 'reluctant new omnivore' (Milburn forthcoming). I have no desire to eat animal products and I am drawn to the moral case for veganism, but I worry that the moral case for eating animal products—in our current world, in some cases—is compelling (Milburn forthcoming; cf. Milburn and Bobier 2022).

alluding to here than on the political questions that have been this book's focus (see, e.g., Lamey 2019; Fischer 2020; Milburn and Bobier 2022).

Consequently, were someone to read this book and adopt a diet that was vegan *but for* oysters, cultivated milk, and the eggs of rescued chickens, I would have no quarrel with them. Indeed, I allow that *they* might have a quarrel with *me*, insofar as I am (they might argue) favouring harmful plant-based sources of protein over beneficial animal-based sources of protein,[11] I am failing to support the first stages of a cellular agriculture industry,[12] and I am failing to rescue chickens in need. There are tricky questions here.

On the other hand, I think I *would* have a quarrel with someone who read this book and subsequently adopted a diet heavy in slaughter-based meat, even while hoping for a future like the one I have pointed towards.[13] It is hard to see how (unnecessarily) supporting the slaughter-based meat industry could be compatible with animal rights today, or how eating slaughter-based meat today could be compatible with hoping for a future in which it is completely banned.

I may be wrong about this. But I am pretty sure I'm not. Support for the ideal championed here, I believe, must surely mean veganism, or something close to it, today—at least for those with control over their diets.[14]

The preceding paragraphs may make for frustrating reading. I might as well have simply said *it's complicated.* So let me be more concrete—but perhaps, for readability's sake, less argumentatively rigorous.

Most existing animal food industries violate animal rights on a massive scale, and so we should not support them. Veganism is thus a natural choice.

But it might not be the *only* choice. There are rights-respecting animal-based (or, to coin a phrase, *sort of* animal-based) foods that some of us have access to today. Among others, these might include plant-based meats, plant-based dairy, and plant-based eggs; the products of precision fermentation; eggs from well-treated companion chickens; and the flesh of non-sentient animals[15] like oysters and jellyfish.

Individuals today may even have access to rights-respecting animal-based foods beyond those I have explored. Take, for example, animal-based foods that would otherwise go to waste. Leftovers from roommates or work meetings, or unwanted surprise gifts or 'extras' on a takeaway order, provide relatively mundane examples. Food 'skipped'—taken from supermarket bins—or

---

[11] Oyster farming, as noted, has compelling environmental credentials.
[12] Although, at time of writing, cultivated milk is not available in the UK, where I live.
[13] I am not speaking about people who eat meat out of necessity.
[14] Most people reading this book, I suspect, are in this position, even if a dietary transition may be *more* challenging for some than others.
[15] Animals that we are basically certain are not sentient.

even roadkill might provide more surprising examples. I am not going to delve into the ethics of 'freegan' eating here (but see Milburn and Fischer 2021; Milburn forthcoming). I simply acknowledge that it is *possible* that a rights-respecting diet here and now might contain animal products that would *not* be accessible in the zoopolis.

Without doubt, there is room for moral debate about today eating non-sentient animals, eggs from backyard chickens, plant-based meat, cultivated milk, roadkill, and so on. I am, nonetheless, inclined to think that it is sometimes (often?) permissible, and may sometimes be *optimal*, to eat these things.

But why might it be optimal? The answer to this question is going to depend upon circumstances. Sometimes, at least in principle, the optimality of eating these foods—the possible fact that eating them is morally preferable to maintaining strict veganism—will tie to the kinds of considerations that have recurred throughout this book. Eating rights-respecting non-vegan foods may help preserve valued rituals; allow us to champion marginalized humans; allow us to make food choices that are only minimally impactful upon sentient animals; help us provide a home for animals in need; and so on.

But even for the purposes of effective activism, a non-vegan diet might be optimal. In our distressingly non-vegan world, a little dietary flexibility might help activists get to the table, might help those in power take advocates a little more seriously, might help make animal-friendly living a little more accessible. These are not reasons to favour rights-violating foods over non-rights-violating foods. But they might be reasons for favouring animal products over vegan products (compare Milburn, 2022c). Of course, working out when (if) these pragmatic considerations apply is tricky.

My point is simply that, in our non-ideal world, certain (limited, unusual) forms of animal-based eating might stretch beyond the permissible to the preferable—or even the obligatory. This is not a conclusion I am personally drawn to. But it is the one I have reached.

Societies

What do the arguments of this book mean for societies attempting to move from unjust to just structures? To be clear, ideal theory, alone and un-supplemented, is not meant to be action-guiding. I have offered indications of what the food system of the zoopolis *might* look like. But even if I had been firmer, saying that such-and-such a food system was what justice *demands*, what that means for non-ideal theory remains open. Pertinently, states probably would not be obliged to switch to that alternative food system tomorrow. This is for at least two reasons.

First, this may be impossibly impractical. Many people may not support that alternative food system; the infrastructure (technology, oversight, laws, etc.) of that alternative food system may not exist; and so on. Thus, at best, the conception of justice would demand that states begin to develop these legal structures, begin to push for this technological advancement, begin to shift the expectations of the populace, and so on.

Second, the switch itself raises ethical questions that I have not explored here. Although one of the advantages of the kind of food system I point towards is that it leaves fewer people 'behind' than would a whole-food plant-based system, a shift away from slaughter-based animal industries will still adversely impact many people.

For example, people who currently make a living from unjust forms of animal agriculture (fishing, hunting ...) will lose their investments (money, time, the planning of the shape of their lives ...). What, if anything, are they owed? Yes, they may be able to find new, *good*, work in the zoopolis. (That may even be work that does not look *too* different from their old work.) And yes, they may still be able to access the foods valuable to them. But let us be clear that the impact upon them will likely be substantial and negative.

There are a range of possible answers. Maybe the state should simply tell them that they can no longer do what they were doing, and that they must find new ways to live. Or maybe they could be 'grandfathered in', or at least warned that what they are doing will become illegal, allowing them to continue their careers *for now*. Maybe the state should offer them ample support in retraining and seeking new employment, meaning that they lose their investments, but their lives are far from ruined. Or maybe the state should 'buy them out', meaning they lose a way of life, but do not lose their (financial) investments. Or the state could combine these strategies.

All of these possibilities come with advantages and disadvantages. A hard-line approach—simply telling pastoral farmers (and similar) to find new lives—pushes the burdens of transition onto farmers' shoulders, and that seems unfair. Granted, they are doing something unjust, in farming rights-bearing animals. But we collectively demand their products. Given that, we could surely (so might argue the farmers!) share the burden of the transition more equitably. On the other hand, this hard-line approach has the advantage of showing no support for continued rights violations—they end, now. That is surely its key strength.

'Grandfathering in' current animal farmers and/or giving advanced warning of an impending ban are routes that do not share this advantage. Such approaches might be (or seem) more respectful to current farmers, insofar as they do not throw away the farmers' investments. But these options mean, to put it bluntly, that slaughterhouses remain open; animals continue to have their rights violated for the next few years (or decades). And, yes, maybe liberation is on the horizon. But that is not much consolation for the pig at the abattoir door. Although this is troubling, perhaps—regrettably—options like this are the practical step in some cases.

What about the other options—of the state 'buying out' farms, and/or supporting farmers in their transitions away from animal farming? Well, if combined with a hard-line approach, this has the dual advantage of not placing a disproportionate burden on the shoulders of farmers *and* of offering immediate respite to animals. But that is not to say that these options are without impediments. First, they would be expensive. Second, people might be concerned about the state offering support to those (seen as) primarily responsible for violence against animals. ('She's made a career killing animals, and now that we've finally banned it, she gets early retirement and a payout?') Third, there are myriad practical questions the state will need to address—what other jobs will these people take up? How will the state price the animals/land/equipment? What will the state do with the animals/land/equipment afterwards? What about those people who lose their jobs or lifestyles but do *not* own animals/land/equipment? And so on.

I am not going to try to adjudicate between these options. We need much more work on these questions if the state is to play a major role in the transition from our currently unjust society to a zoopolis. But these questions of 'just transitions' are *not* those most pressing today. The fact is that states are not grappling with the question of how to equitably shift to a food system in which we respect animals' rights. Indeed, in many cases, they are grappling with the question of how to shift to a food system in which we kill *more* animals (e.g. 'sustainable intensification'), or in which we lessen the adverse impacts of killing animals *on humans* (e.g. 'technofixes' making slaughter-based meat greener). Or perhaps they are focused on preventing activists from impacting current practices at all (e.g. 'ag-gag' laws).

In any case, today, non-ideal theory for animal ethicists is more about getting animal rights on the agenda than it is about the routes that states can/should take to realize justice. It is directed, then, primarily at animal activists.

Activists

How could (should) the arguments of this book impact animal activists? Again, as ideal theory, they mean nothing on their own. And, to be clear, there may be more important matters to be addressed in non-ideal theory that are well beyond the scope of the present book—questions about forming alliances with other movements, about getting animals onto party political agenda, about dividing time and resources between legal, social, and political forms of activism, and so on.

Further, this is a book about food systems, and so it says little about the division of activist resources between (say) companion animals, farmed animals, animals used in vivisection, and so on. (Although, for what it is worth, and as indicated earlier, I see the food system as a key area for animal activists. With the possible exception of wild animals, it is animals exploited for food who face the most egregious injustices.) Nonetheless, I here offer some reflections to *start the conversation* about what a non-ideal version of my ideal vision might look like. My focus, of course, is on food systems.

We can start by distinguishing three broad approaches to animal activism. Each is *in principle* compatible with a range of possible visions of future co-relations with animals.

First, we can identify *abolitionist* activism (Francione and Charlton 2015). This opposes single-issue campaigning (say, the anti-fur campaign) or efforts to make animal use more humane (say, campaigning for free-range eggs), and instead calls for vegan education.

Readers might fairly see abolitionist activism as the route to the future sketched in this book. Although I have not advocated a strictly vegan future, I have advocated, in important senses, a *near*-vegan future. I have not championed 'humane' slaughter or 'sustainable' fishing; I have championed the abolition of slaughter and (lethal) fishing. Yes, the future I have envisioned *does* contain institutions reminiscent of farms—whether they are farms housing non-sentient invertebrates, or the workplaces of animal labourers. But these institutions (we might think) are *so* different from most contemporary farms that the appropriate activist practice would be to *abolish* contemporary farms and 'start again'.

This abolitionist approach might be pragmatic. It might be easier to remove existing institutions and then design new ones than attempt to transition current institutions into something new. It might also be ethical, as it does not call on us to tolerate unjust practices and relations. This dual focus on pragmatism and ethical viability is at the core of non-ideal theory (Garner 2013, chapter 1).

Let us look to the second form of activism, one typically seen as abolitionism's key competitor (Francione and Garner 2010). Welfarists champion transition towards a just future through piecemeal legislative and practical change, aiming to reduce animal suffering where they can. Thus, they might target putatively low-hanging fruit (like bloodsports) rather than the behemoth that is the food system. When challenging the food system, meanwhile, they will focus on the (supposedly) comparatively achievable goals of making it a little less unjust through, say, championing more 'humane' slaughter.

Again, some readers might associate the vision I have advocated with welfarism. My discussion of invertebrate farming focused on suffering, while my discussions of cellular agriculture and eggs called for suffering-free animal agriculture. Described like this, it sounds like my vision is one of a welfarist utopia. (This is far from how I see my conclusions—but I accept that it is probably how some others will view it.) And how does one reach a welfarist utopia? Well, presumably by pushing back against suffering in contemporary animal agriculture, gradually turning our current farms into places free from suffering.

New approaches to animal rights find a natural ally in a third approach to animal activism. This approach focuses on animals' membership rights, championing moves to recognize animals as belonging to certain (traditionally human) categories: family members, workers, citizens, and so on. As the law recognizes animals as members of these groups, the case goes, they can achieve the protections and rights associated with said membership (Kymlicka 2017).

The food system I have sketched draws strongly upon new visions of animal rights. And, indeed, it depends upon the *particular* rights that animals are entitled to as members of given social groups—workers especially. As such, it may seem that the social membership approach to animal activism complements the vision I have forwarded. Perhaps the non-ideal approach naturally suggested by the ideal I have championed is campaigning to recognize animals as citizens, as workers, as society members, and more so that they can come to play a part in our food systems not as mere livestock, but as co-contributors.

I do not necessarily align myself with any of these three approaches. Equally, I am not convinced that a non-ideal complement to the ideal theoretic considerations I have developed will necessarily fall neatly at a particular place on the abolitionism/welfarism/membership trichotomy.

I propose the following. Taking steps towards the future envisioned in this book will involve advocating food system change that is not solely about increasing (say) the farming of legumes and pulses and decreasing (say) industrial animal agriculture. It will involve offering *some* support to *some* forms of meat, milk, and egg production. But what this means requires paying close attention to the *particular* food industries that (may) make up a part of the ideal food system, and reflecting on their relationship to extant food industries.

Take invertebrate farming. Activists should be championing sentience as the foundation of animal welfare (*rights*) law, as well as championing expansions of legal definitions of sentience to capture sentient animals excluded. But, conversely, they should be cautiously placing *non*-sentient animals on the agenda as a possible answer to questions about the future of food. They must be cautious for at least two reasons.

First, insofar as some sentient animals have been (and are) assumed non-sentient, the inattentive may interpret the championing of farming of 'non-sentient' animals as support for (say) fish farming. And even if people do acknowledge that fish are sentient, they may argue (in my view, incoherently) that fish are 'less' sentient than terrestrial vertebrates, and their farming is thus unproblematic, or less problematic.

Second, there is the risk that championing the farming of (non-sentient) invertebrates will shore up the farming of vertebrates, insofar as the animal feed industry is a major source of income for invertebrate farms (Sebo and Schukraft 2021). My guess is that one way around this is offering *specific* proposals. Perhaps activists should champion *oyster* or *jellyfish* farming *for human consumption*, rather than simply 'invertebrate farming'.

Animals who *may* be sentient are trickier. On the one hand, it might seem obvious that animal activists should support the search for sentience in these beings. However, they should also be careful about (apparently or actually) supporting (potentially) harmful practices of vivisection. (In practice, much research on invertebrate sentience involves the literal torture of animals— or it does if the animals in question *are* sentient.) In the public sphere, it is important that uncertainty is stressed. Animal activists lose credibility if they appear to rest their arguments upon questionable claims of certainty.

That said, I doubt that it is wise for *activists* to champion the consumption of these animals *today*. While I do not think that the balance of risk should fall entirely on the heads of those who would benefit from farming these animals, I do think that these animals need their advocates.

Plant-based meats, I think, are an easier proposition. (Let us include plant-based milks, eggs, honey, and so on under the same banner.) In my view, these are foods that animal activists should champion, even if activists should be quick to concede that plant-based meats are frequently junk food. These products are not harmful to animals, and although I will be the first to allow that they face legitimate ethical challenges, I believe we can overcome these challenges. (In Chapter 3, I considered challenges to these products in the light of ideal theory—here is not the place for exploring those challenges in non-ideal theory, but I believe them far from decisive.)

There are, however, interesting questions about how to balance two apparently conflicting claims. First, animal activists should want to stress that this is *meat*—it is (in principle[16]) able to fill the culinary gap left by slaughter-based meat. But second, they should want to stress that it is *different* from slaughter-based meat.

Many activists will have encountered challenges such as 'If you want to eat a burger, just eat meat!' or 'If you want to eat vegetables, just eat a salad!' To put it bluntly, these are so wide of the mark that it is tricky to unpack them. But that does not stop thoughts like that being a genuine impediment to progress. Activists championing plant-based meat need to find a balance—and I do not, here, have specific guidance about how to do that, even if I do believe that activists *should* be championing plant-based meats as a route to a more just future.[17]

Cellular agriculture is, I think, the area that it is *most* important for activists to get right. It offers (in my view) considerable potential to remedy current injustice, and represents a major change for the global food system between the one we have currently and the one envisaged in this book. Activists should throw themselves behind cellular agriculture, but be ready to challenge it when it takes a wrong step. An easy example is the use of FBS. In my view, activists should *not* champion FBS, and they should use whatever influence they can get within the cellular agriculture industry to challenge it. It is also important that animal activists speak up in defence of the animals who (might) continue to be used by the cellular agriculture industry as 'donors'. Even today, in cellular agriculture's early days, we should not be tolerating cellular agriculturalists finding 'donors' in the slaughterhouse.

To the broader public, legislatures, and businesses, however, cellular agriculture should be championed as a way to replace animal products with humane alternatives. As with plant-based meat, activists must perform a balancing act. On the one hand, this is *real* meat (or milk, or what-have-you). On the other, this is *different* from slaughter-based meat (or milk, or what-have-you).

---

[16] Someone might fairly claim that plant-based meats are not as good as slaughter-based meats. That contention becomes less convincing by the year.

[17] The alternative view is that we cannot successfully negotiate this balancing act, meaning activists should not champion *anything* meaty, to decentre meat. As I have argued, we do not need to decentre meat. I aspire to a future in which there *is* meat.

Perhaps the trickiest question about contemporary activism arises when it comes to the possibility of genuinely humane forms of egg (and milk) farming. On the face of it, the non-ideal side of the coin might seem to be the pursuit of welfare legislation, so that we can slowly turn today's ostensibly humane forms of farming into *genuinely* humane forms of farming. But I am not sympathetic to this idea. I am not sure that the kind of genuinely humane egg farming I have explored is something realizable soon.

Rather than actively campaigning for it, I suggest that the real message for activists is to look to (micro)sanctuaries as places where advocates can experiment in different ways of co-living with animals (Donaldson and Kymlicka 2015). Alongside this, activists should pursue legal reform to recognize animals as belonging to relevant membership groups, bringing them the protection afforded to members of these groups. My proposal for humane egg farms is impotent without the possibility of state recognition of animals as workers.

So, I do not see the seeds of tomorrow's egg 'farms' in contemporary egg farming. Instead, I see them in contemporary microsanctuaries—in those people who rescue chickens and eat some of their eggs. Does this mean I think we should, in this non-ideal world, *encourage* people to rescue chickens and eat their eggs? Not quite. But I do think we should encourage people to rescue chickens, and if that means near-vegans eating more eggs, so be it. Or, if the prospect of egg-eating leads to animal advocates rescuing more chickens, all the better for the chickens (cf. Fischer and Milburn 2019). As well as encouraging animal rescue, activists should be open to alternative ways of living with animals, even if they might involve some practices traditionally associated with animal abuse.

Let me be provocative. Activists have often championed the prospect of turning farms into sanctuaries. I do, too. But, in time, there may be room for turning *sanctuaries* into *farms*.

But we must not let any of what I have said undermine the importance of two goals that—thankfully—many activists already place front and centre. They are, first, critiquing slaughter-based meat, milk, eggs, and more, whether this is via abolitionist, welfarist, or membership-based activism. And, second, championing plant-based agriculture.[18] Even if I think that animal activists should be more open to (say) oyster farming and cultivated milk than they sometimes are, critique of paradigmatic forms of animal agriculture

---

[18] Incidentally, contemporary forms of arable agriculture themselves have room to improve (cf. Milburn 2022b, chapter 5).

should remain at, or at least *close to*, the centre of animal activism. And even if I think that there is room for reimagined pastoral agriculture in the zoopolis, that does not change the fact that *arable* agriculture forms the foundation and backbone of the food system of the zoopolis.

Concluding remarks

In this chapter, I have stepped away from questions about specific foods to address some familiar subjects for political theorists: questions about the limits and reach of state action and questions about transitions to just society. I have explained what it means for a state to *permit* a certain food industry, and—more controversially—what it might mean for a state to *support* it.

These reflections, however, have been necessarily incomplete. This is for two reasons. First—echoing a recurring theme—some of these decisions depend upon claims that stretch beyond philosophy, including claims about data that we simply do not have (for example, the number of animals killed in different plant-production methods). Second, I have refrained from committing too strongly to a particular 'flavour' of liberal politics. Instead, I have aimed to indicate the *range* of intellectual resources that liberal theorists could draw upon to justify support for non-vegan food systems.

In addition, I have stepped away from this book's ideal theoretic lens to consider non-ideal theory, exploring what this book's arguments (might) mean for us here and now. My conclusions have been both indicative and speculative, but I have suggested that my arguments might justify certain very specific non-vegan diets today, and that they might legitimately draw the attention of animal activists in curious new directions.

This is the book's penultimate chapter. It is thus appropriate that its conclusions are a little more open—a little less firm—than those of preceding chapters. Much work remains to be done. I have pointed towards a possible future, a particular destination, sketching a picture of what some parts of that future might look like. But my sketching, quite by design, has not amounted to a full picture of that destination. Nor have I fully mapped our route to it.

I nonetheless hope that what I have provided gives readers enough hope that this future is somewhere we *could* go, and somewhere we might *want* to go, that they will join me on this journey. If so, I invite them to help me paint a fuller picture of this envisioned future, map out our route to it, and—of course—move us in the right direction on our collective journey.

Conclusion

Having our cow, and eating her too

In this book, I have explored why—even in the animal-rights respecting state—we should pause before cheering for plant-based food systems. Specifically, I have argued that there are reasons to think the zoopolis (the animal-rights-respecting state) should *permit* and even *champion* food systems containing elements beyond plant-based whole foods. The diet of zoopolitans might include certain invertebrates, or invertebrate products; plant-based meats, milks, and eggs; the products of cellular agriculture; or even products from animal workers on genuinely humane farms. And my focus is not just on individual diets—industries producing these foods may well survive (and thrive) in the zoopolis.

This conclusion is bold, but it could be bolder. I have stopped short of arguing that the all-things-considered *best* food system, in the zoopolis or elsewhere, contains the foods explored. To put it another way, although this is a work of ideal theory, it has not sought to offer a blueprint of the perfect system, in the tradition of (say) Plato's *Republic*. This is for two reasons.

First, calculating the most just food system is too complex for a single book. The calculation must account for any number of competing factors. This book, of course, has focused on animal rights, with nods towards (inter-human) food justice, environmental sustainability, world hunger, and more. But it has said little about other concerns, such as property rights (including intellectual property rights); food safety; social welfare; (human) workers' rights; and more.

Indeed, even when it comes to animal rights, there are difficult questions I have not answered. For example, although I have pointed towards how certain animal-based food production methods could involve less animal harm than some forms of arable agriculture, I have not delved into the numbers—indeed, how could I, given that the numbers are not known (Fischer and Lamey 2018)? Assorted experts need to do more work to detail how much harm farmers inflict on animals in different production methods. (And, do not forget, not all harms are unjust.) And, for reasons of space, I have said almost nothing about feeding animals themselves, which must be part of the conversation about respectful food systems (cf. Milburn 2022b).

But, second, the practical conclusion of my argument is not only a question for political philosophy. For example, I have argued that people's conceptions of the good should influence our food system designs. Questions about diverse conceptions of the good life depend upon what conceptions of the good life people *actually have*. While we can do much to reflect on the conceptions of the good life that people have today, we do not know what conceptions will prevail tomorrow. Will people continue to value the tastes and foods they value today? Will they continue to value the jobs and lifestyles they value today? Will they continue to express their cultural and religious identities as they do today? Indeed, will they even have the same religious and cultural identities that they have today?

What is more, liberal political theories are (almost?) invariably *democratic* theories. Questions about how the state prioritizes competing demands and allocates scarce resources are frequently democratic questions as well as (or instead of) questions of justice. And political theorists have no business declaring what the outcomes of democratic or deliberative fora will be—even if they can offer insights in how to design said fora. If there is no democratic will for the state to (say) support the cellular agriculture industry, then, perhaps, so be it.

That said, we must be clear about the limits of democratic will. It cannot mean, for example, that injustice is no issue. Limited democratic will to prevent the subjugation of a minority group is not a good reason for the state to champion or tolerate said subjugation. (And that is as true of animals *qua* minority group as, say, racial minorities.) But lack of democratic will to support an ailing industry might be a good reason to let the industry fade away—even if that might mean some ways of life are no more.

Nonetheless, I hope to have offered considerable food for thought concerning future food systems, and hope to have changed (or, minimally, challenged) what animal rights *means*. My conclusions are, I accept, potentially troubling. At the time of writing, I am a committed vegan. And, naturally, I long assumed that animal rights meant veganism. Initially, as I started to think otherwise, I was uncomfortable with my own conclusions. But I did not reach this book's conclusions lightly. I hope it is clear that I do not offer these conclusions as a contrarian or provocateur—nor do I intend the arguments to be a complicated *reductio*. They are the result of serious research and reflection on how we should live together—and serious research and reflection can lead to conclusions that are both alarming *and* correct.

In this conclusion, I summarize the book's arguments, before asking whether I (or others) could take them further. I end with some brief comments on getting questions about food systems right, and getting them wrong.

Animal rights without veganism? A summary of arguments

If we are envisaging a just food system while holding animal rights in mind, we are likely to imagine a future with no animal products—no meat, milk, eggs, honey, or similar. This would be a future without the egregious rights-violations of animal agriculture, hunting, and fishing. And it would be a future with lower levels of many *human* harms associated with animal agriculture. But we should not be too quick to embrace a food system made up of only plant-based whole foods.

Why? For a start, people have diverse conceptions of the good. They have a range of different ideas about what the good, *meaningful* life is. Many of these conceptions involve access to animal-based foods, or things (jobs, landscapes, cultures, and more) associated with the production of animal-based foods. Plant-based food systems would seemingly deny these people the opportunity to pursue their good.

The existence of these diverse conceptions does not demonstrate that plant-based food systems are incompatible with the demands of justice, or that the most just food system would not be plant-based. But it does demonstrate that we have good reason to be open to the possibility of a system incorporating animal products. Liberal states have little business preventing that which is compatible with justice—and we should heed the warning from critics of animal rights that we cannot allow concern for animals to become an excuse for the state to run our lives for us (Machan 2004, 23). And, further, liberal states may have reason to create space (insofar as the demands of justice allow) for people to realize their diverse conceptions of the good.

Some considerations do point towards the *possibility* (recall the enormity of these questions) that a plant-based food system may be incompatible with the demands of justice, or at least may not be the most just conceivable system. First, animal-based foods may be important in the realization of traditional goals of food justice, such as food sovereignty and food security. (For example, cultivated meat, as a nutrient-dense food, could be valuable for those in need of food aid.) Second, due to harms to animals in arable agriculture, it is possible that the production of some animal-based foods is less harmful than the production of some plant products. (For example, sustainable oyster-farming, assuming oysters are non-sentient, is likely much more animal-friendly than intensive soy farming.) And, finally—although the significance of this is open to discussion[1]—plant-based food systems face a curious challenge in that they struggle to find a *place* for the animals they are saving. (For example, a future in which chickens live out their lives on genuinely humane egg farms might seem more chicken-friendly than one in which they are all but extinct.)

If the zoopolis has reason to permit and even endorse a rights-respecting, but animal-based, food system, the natural next question is what such a food system might look like.

It will likely contain the bodies and products of non-sentient invertebrates. Non-sentient animals lack rights, meaning humans may freely farm or gather them. The state has no business *preventing* these things. (At least, not for the animals' own sakes.) Insofar as their farming or gathering may help people realize their conceptions of the good, may help alleviate food-justice concerns, and may help other animals, these are things that the state should perhaps be *supporting* and *endorsing*.

The state, however, faces difficult questions around the regulation of those industries utilizing animals who (or that) *may* be sentient. While those animals that are vanishingly unlikely to be sentient should be considered fair game, and those probably sentient should be protected as if they are, this leaves another group of animals who are *probably* not sentient, but may be. These categories are hazy. Example taxa may be jellyfish in the first category; decapod crustaceans in the second; and insects in the third.

I have proposed that we should translate the *possible* rights of members of this third group into a single *legal* right: A right against the infliction of suffering. (A right against treatment that, our best estimates suggest, would cause these animals to suffer *if* they were sentient.) If we can farm these animals in a way consistent with this legal right, the state should not prevent it. Although, given the risks involved (i.e. the risk of the violation of the plausibly sentient animals' justice-based rights), it is unlikely that this is the sort of animal agriculture that the state should *encourage*.

Decisions like this, however, are a balancing act. If it is the case that these forms of farming are *particularly* friendly to other animals and the environment, *particularly* important for realizing the goals of food justice, or even

---

[1] This observation does not mean plant-based food systems are unjust, or that they do not respect animal rights. But, intuitively, liberating animals should not mean their all-but extinction. Indeed, this intuition has been one central to the shift from the old abolitionist approaches to animal rights to new political conceptions of animal rights.

(perhaps) *highly* important for many people's conceptions of the good life, then things may be different.

For example, it is at least *plausible* that insect farming is more environmentally and animal-friendly than arable agriculture (Meyers 2013; Fischer 2016a), and thus that the zoopolis could buy insect foods for public institutions even while insects *might* be sentient. It is at least *plausible* that the gathering of marine invertebrates could be particularly important for the food security/sovereignty of marginalized coastal communities, and thus that the zoopolis could allocate protected foraging areas even while these invertebrates *might* be sentient. It is at least *plausible* that beekeeping as a way of life and honey as a valued food are sufficiently important for people that public money could support apicultural education, even while bees *might* be sentient. In these kinds of cases, perhaps, on balance, the state should support industries or lifestyles even despite the moral risk involved.

The zoopolis is on firmer ground when it comes to plant-based meats, milks, eggs, and so forth. While critics might pooh-pooh these foods as not *real* animal products, they nonetheless face critique from animal activists who instead advocate plant-based whole foods. I have argued that neither these challenges, nor challenges from those who worry about 'processed' foods, should prevent the zoopolis from permitting their production.

Perhaps plant-based meats (and similar) will not always be the *healthiest* foods, or the most *environmentally friendly* foods—even if (typically) at least as healthy as, and more environmentally friendly than, comparable slaughter-based animal products. Nonetheless, they are, on balance, something that the zoopolis might well champion, given the importance that these foods can have in people's lives. If it just is not (to pick some examples from my own cultural background) Christmas dinner without turkey, watching the game with dad without a meat-and-potato pie, or a day at the seaside with grandkids without ice cream, then it is all the better that (plant-based) turkey, pies, and ice cream remain.

Many of the considerations applied to plant-based animal products apply, too, to cultivated animal products. They, too, face challenges both from animal activists for replicating 'real' animal products and from apologists for animal agriculture for being (in some sense) bad food. But there are other challenges, too. I have argued that the abuse of animals in the development of these products should not lead us to condemn the products themselves if they are made respectfully, and that the ethical questions faced by the particular technologies of cellular agriculture (e.g. precision fermentation) are not too worrying. But cellular agriculture (especially cultivated meat) also faces questions about its fit in the broader food system, including about the role of *animals*.

Cultivated meat needs animal cells. I have endorsed two existing models by which producers could source animal cells, and proposed a third. The zoopolitical vision of a 'pig in the backyard' as a source of cells captures the imagination of animal activists open to cultivated meat; the techno-futurist model of 'mail-order cells' captures the imagination of the engineers drawn to cellular agriculture. But we could complement these visions, I think, with a recognizably *agrarian* vision of animals living out their lives on farms—but farms on which animals are colleagues of farmers, not mere livestock. These animals would be protected by rights as animals, as community members, and (most crucially) as workers.

But if we could envision respectful farmers keeping animals for their cells, could we imagine respectful farmers keeping animals for other products? Meat is off the table—lethal workplaces are not respectful. But we could envision respectful egg farms, and perhaps dairy farms. I sketched a respectful egg farm, but leave the possibility of a respectful dairy farm for others to explore. Now, the observation that egg and dairy farming *could* be respectful does nothing to exonerate the egg or dairy industries of today. And the envisioned agriculture would not produce *cheap* eggs or dairy—most eggs and dairy in the zoopolis would be plant-based or cultivated, or perhaps 'home-grown'. But it does provide a vision of a home for chickens and cows, an agrarian lifestyle for those who value it, and a speciality food for those who desire it.

Taking the argument further

Other foods

Could there be other sources of animal-based products that could find their home in the zoopolis? Recently, literatures in ethics around 'unusual eating' (Fischer 2020) and 'new omnivorism' (Milburn and Bobier 2022) have emerged which chart assorted harm-free (or comparatively harm-light) sources of animal-based foods.[2] Much of this discussion has focused on 'waste' animal products, like roadkill, or animal products that would otherwise end up in landfill.

---

[2] As I define these terms (Milburn and Bobier 2022), unusual eating is about identifying 'unusual' animal products compatible with animal rights, while 'new omnivorism' is about defending particular animal products on the grounds that their consumption would minimize physical harms to animals (e.g. animals killed incidentally in arable agriculture).

On the face of it, these sources of animal-based foods are more useful for intrepid near-vegans than for states contemplating food systems. But this need not be the case. The zoopolis has reasons to minimize food waste, which may include systems for the redistribution of edible food—plant-based or otherwise. Arable agriculture is harmful to animals, and an efficient food system will involve *less* arable agriculture, and thus (all else equal) less harm to animals.

Similar *could* be true about systems redistributing edible roadkill—which already exist in some parts of the world.[3] The redistribution of roadkill raises a host of other questions, such as the question of whether we have a duty to respect the corpses of animals killed on the road.[4] And none of this should distract from the fact that we should be doing more to prevent animal deaths on the road. But there is at least an open question about whether 'waste' animal products might be a legitimate source of food in the zoopolis.

Consider another possibility. In Chapter 6, I mentioned collecting wild animals' eggs in passing. There are the eggs of wild birds, for instance. The overwhelming majority of birds' eggs are edible, although some are more favoured than others; licensed professionals harvest gull eggs in the UK, for example. There are the eggs of wild fish; roe is both a delicacy beloved by foodies and a traditional food of many peoples. There are the eggs of wild reptiles, including turtles, crocodiles, and alligators. And there are even the eggs of insects. Ant eggs (and larvae) appear in East Asian and Mexican dishes, for example—'Mexican Caviar' does not come from fish. (People farm some of these animals for eggs, but not all of them.)

Would harvesting these wild animals' eggs involve the violation of animals' rights? This raises questions worth exploring. For example, the harvesting of gulls' eggs might violate the rights of gulls themselves—they are, after all, minimally attentive parents. Sea turtles, meanwhile, abandon their eggs after laying them. On the other hand, the harvesting of sea turtle eggs raises concerns among conservationists. But, to my knowledge, there is no environmentalist outcry over the gathering of black-headed gull eggs on the British coast. I leave it as an open question whether there would be room for the harvesting of wild animals' eggs in the zoopolis—although I think the question a very interesting one.

---

[3] In Alaska, roadkill belongs to the state, which redistributes it to charitable groups (Abbate 2019, 171).
[4] In Chapter 5, I pointed towards a duty to respect the bodies of animals as a reason to oppose corpse-farming. But that argument depended upon farmed animals being members of our community, which animals killed on the road may not be. Perhaps, then, corpse-farming is impermissible even while the eating of roadkill is permissible. Compare Milburn and Fischer 2021 and Milburn 2020.

Beyond food

I could extend the argument of this book by looking to systems *beyond* food. For example, I could imagine a similar argument about clothing. Lots of people value having access to leather, wool, fur, silk, and so on. They work with it; dressing in a certain way is important for their cultural identity; they like to look certain ways; and so forth. (And there may be other reasons to worry about an animal free global 'clothing system'—are animals harmed in cotton harvesting? Does 'clothing justice' favour certain communities continuing to have access to wool? Could a rights-respecting mohair industry provide a 'place' for goats? And so on.)

Animals have rights, and so these concerns should not lead us to say that (for example) the fur industry is anything but unjust. But they do give us reason to hope that we could have a rights-respecting, but non-vegan, clothing system—that we could have our cow and *wear* her too. We could champion a clothing system containing products derived from non-sentient animals (pearls?)—or even some fabrics from animals who *may* be sentient (silk?). We could champion the development of 'plant-based' leather. We could champion the use of cellular agriculture to develop cultivated fabrics. (California-based Bolt Threads makes cultivated spider silk. Amsterdam-based Furoid is developing cultivated fur and wool. Modern Meadow and Virtolabs, both American, are working on cultivated leather.) And we could even create space for animals protected by workers' rights as a source of *some* raw materials for clothing. Farming sheep and goats for wool? Perhaps feasible. Farming cattle for leather or mink for fur? Perhaps not.

And why stop at food and fabrics? Anywhere we use animal products today, we can ask these same questions. And—to the frustration of vegans—animal products are *everywhere*. Medicine. Beauty products. Cosmetics. Cleaners. Building materials. Vehicles (and roads). Money—*literal banknotes*. Ink, for tattoos, books, packaging. Toys, games, equipment used in sports. In each case, we can ask the following: Would there be a significant loss if we stopped using animal products? (In banknotes, no—who cares? For cricket balls, given the enthusiasm some people unaccountably have about cricket, yes.) If there would: Is there a way to provide the animal product sought—or preserve the benefit sought—while respecting animals' rights? If not, then, regrettably, we must bear that loss. If so, then we have a good (although defeasible) reason to create a respectful society able to preserve that benefit—even if it means developing new technologies, practices, or interspecies relationships.

This book provides indications for how we might be able to answer these questions—but there are many more questions we could ask, and many answers we are yet to imagine. Respecting both animals and humans is, in a word, *hard*.

Getting it right, getting it wrong

In this book, I have said some surprising things; some things that will, I suspect, provoke hostile responses from other animal ethicists and animal activists. If what I say is wrong, I welcome correction. But I hope advocates of animal rights do not dismiss my work simply because I endorse something other than veganism.

I accept that apologists of animal agriculture could hold up my conclusions or arguments (ignorantly or disingenuously) as a defence of their own dietary habits, or their own favoured disrespectful food system. This is a risk, and one that is inevitable when presenting any but the most straightforward claims. But I hope that I have been sufficiently clear with my reasons, and sufficiently clear with my support for animal rights, that anyone who takes even a short time to review my claims will see that they are decidedly not a justification of the status quo, or anything close to it.

Indeed, anyone who has reached these closing lines will know that I am a fierce critic of the status quo. In its place, I have offered a vision of what our future food system could look like. Food is worth talking about because it is the area in which animals face the most egregious harms at human hands—it is, I believe, the space in which animals face the most significant injustices. Indeed, this dry claim fails to capture the hell that is animal agriculture. What we do to animals in pursuit of food certainly is unjust, but simply labelling it as such feels like an understatement. It is terrible. It is awful. It is evil. It is wrong. It is sickening ... I am not sure I have the words.

I want to see our relationship to animals change, quite drastically. I believe that the direction in which I have pointed in this book is not only the *right* direction, but is one in which many people, from the current food system's harshest critics to its most ardent apologists, could see some merit. Stronger, I think it is something that *many* of us—from vegans to meat eaters—could get behind. If that is right, then maybe we, collectively, could have the will to make a change for the better, and start on the long journey to overturn something about which I can barely overstate the wrong, the injustice, the *horror*. What could be more worthwhile?

Bibliography

Abbate, Cheryl. 2019. 'Save the Meat for Cats: Why It's Wrong to Eat Roadkill'. *Journal of Agricultural and Environmental Ethics* 32 (1): pp. 165–182. doi: 10.1007/s10806-019-09763-6.
Abbate, Cheryl. 2020. 'Animal Rights and the Duty to Harm: When to be a Harm Causing Deontologist'. *Zeitschrift für Ethik und Moralphilosophie* 3: pp. 5–26. doi: 10.1007/s42048-020-00059-3.
Abrell, Elan. 2021. 'From Livestock to Cell-Stock: Farmed Animal Obsolescence and the Politics of Resemblance'. *Tsantsa* 26: pp. 37–50. doi: 10.36950/tsantsa.2021.26.6.
Adamo, Shelley Anne. 2016. 'Do Insects Feel Pain? A Question at the Intersection of Animal Behaviour, Philosophy and Robotics'. *Animal Behaviour* 118: pp. 75–79. doi: 10.1016/j.anbehav.2016.05.005.
Adams, Carol J. 1990. *The Sexual Politics of Meat: A Feminist-Vegetarian Critical Theory*. London: Bloomsbury.
Adams, Carol J. 2016. 'Ethical Spectacles and Seitan-Making: Beyond the Sexual Politics of Meat—A Response to Sinclair'. In *The Future of Meat Without Animals*, edited by Brianne Donaldson and Christopher Carter, pp. 249–256. London: Rowman & Littlefield International.
Adams, Carol J. 2017. 'Feminized Protein: Meaning, Representations, and Implications'. In *Making Milk: The Past, Present and Future of Our Primary Food*, edited by Mathilde Cohen and Yoriko Otomo, pp. 19–40. London: Bloomsbury Academic.
Adams, Carol J., and Matthew Calarco. 2016. 'Derrida and *The Sexual Politics of Meat*'. In *Meat Culture*, edited by Annie Potts, pp. 31–53. Leiden: Brill.
Ahlhaus, Svenja, and Peter Niesen. 2015. 'What is Animal Politics? Outline of a New Research Agenda'. *Historical Social Research* 40 (4): pp. 7–31. doi: 10.12759/hsr.40.2015.4.7-31.
Alkon, Alison Hope. 2012. 'Food Justice: An Overview'. In *Routledge International Handbook of Food Studies*, edited by Ken Albala, pp. 295–305. Abingdon: Routledge.
Alkon, Alison Hope, and Julian Agyeman, eds. 2011. *Cultivating Food Justice: Race, Class, and Sustainability*. Cambridge, MA: The MIT Press.
Alvaro, Carlo. 2019. 'Lab-Grown Meat and Veganism: A Virtue-Oriented Perspective'. *Journal of Agricultural and Environmental Ethics* 32 (1): pp. 127–141. doi: 10.1007/s10806-019-09759-2.
Alvaro, Carlo. 2020. *Raw Veganism: The Philosophy of The Human Diet*. Abingdon: Routledge.
Ankeny, Rachel A., and Heather J. Bray. 2018. 'Genetically Modified Food'. In *The Oxford Handbook of Food Ethics*, edited by Anne Barnhill, Mark Budolfsen, and Tyler Doggett, pp. 95–111. Oxford: Oxford University Press. doi: 10.1093/oxfordhb/9780199372263.013.40.
Archer, Mike. 2011a. 'Ordering the Vegetarian Meal? There's More Animal Blood on Your Hands'. The Conversation. https://theconversation.com/ordering-the-vegetarian-meal-theres-more-animal-blood-on-your-hands–4659.
Archer, Michael. 2011b. 'Slaughter of the Singing Sentients: Measuring the Morality of Eating Red Meat'. *Australian Zoologist* 35 (4): pp. 979–982. doi: doi.org/10.7882/AZ.2011.051.
Barnhill, Anne, and Matteo Bonotti. 2022. *Healthy Eating and Political Philosophy: A Public Reason Approach*. Oxford: Oxford University Press.
Barnhill, Anne, Katherine F. King, Nancy Kass, and Ruth Faden. 2014. 'The Value of Unhealthy Eating and the Ethics of Healthy Eating Policies'. *Kennedy Institute of Ethics Journal* 24 (3): pp. 187–217. doi: 10.1353/ken.2014.0021.
Barry, Brian. 2000. *Culture and Equality: An Egalitarian Critique of Multiculturalism*. London: Polity.
Barry, Christian and David Wiens, 2016, "Benefiting from Wrongdoing and Sustaining Wrongful Harm", *Journal of Moral Philosophy*, 13(5): 530–552. doi:10.1163/17455243-4681052
Basheer, Atia, Chris S. Haley, Andy Law, Dawn Winsor, David Morrice, Richard Talbot, Peter W. Wilson, Peter J. Sharp, and Ian C. Dunn. 2015. 'Genetic Loci Inherited from Hens Lacking Maternal Behaviour

Both Inhibit and Paradoxically Promote this Behaviour'. *Genetics Selection Evolution* 47 (100): pp. 1–10. doi: 10.1186/s12711-015-0180-y.

Belshaw, Christopher. 2015. 'Meat'. In *The Moral Complexities of Eating Meat*, edited by Ben Bramble and Bob Fischer, pp. 9–29. Oxford: Oxford University Press. doi: 10.1093/acprof:oso/9780199353903.001.0001.

Benyshek, Daniel C., Melissa Cheyney, Jennifer Brown, Marit L. Bovbjerg. 2018. 'Placentophagy Among Women Planning Community Births in the United States: Frequency, Rationale, and Associated Neonatal Outcomes'. *Birth* 45 (4): pp. 459–468. doi: 10.1111/birt.12354.

Bernstein, Justin, and Jan Dutkiewicz. 2021. 'A Public Health Ethics Case for Mitigating Zoonotic Disease Risk in Food Production'. *Food Ethics* 6 (9): pp. 1–25. doi: 10.1007/s41055-021-00089-6.

Bhat, Zuhaib F., James D. Morton, Susan L. Mason, Alaa El-Din A. Bekhit, and Hina F. Bhat. 2019. 'Technological, Regulatory, and Ethical Aspects of In Vitro Meat: A Future Slaughter-Free Harvest'. *Comprehensive Reviews in Food Science and Food Safety* 18 (4): pp. 1192–1208. doi: 10.1111/1541-4337.12473.

Birch, Jonathan. 2017a. 'Animal Sentience and the Precautionary Principle'. *Animal Sentience* 16 (1): pp. 1–15. doi: 10.51291/2377-7478.120.

Birch, Jonathan. 2017b. 'Refining the Precautionary Framework'. *Animal Sentience* 16 (20): pp. 1–18. doi: 10.51291/2377-7478.1279.

Birch, Jonathan. 2018. 'Degrees of Sentience?' *Animal Sentience* 21 (11): pp. 1–3. doi: 10.51291/2377-7478.1353.

Blattner, Charlotte E. 2020. 'Animal Labour: Toward a Prohibition of Forced Labour and a Right to Freely Choose One's Work'. In *Animal Labour: A New Frontier of Interspecies Justice*, edited by Charlotte E. Blattner, Kendra Coulter, and Will Kymlicka, pp. 91–115. Oxford: Oxford University Press. doi: 10.1093/oso/9780198846192.003.0005.

Blattner, Charlotte E., Kendra Coulter, and Will Kymlicka, eds. 2020. *Animal Labour: A New Frontier of Interspecies Justice*. Oxford: Oxford University Press.

Blattner, Charlotte E., Sue Donaldson, and Ryan Wilcox. 2020. 'Animal Agency in Community: A Political Multispecies Ethnography of VINE Sanctuary'. *Politics and Animals* 6: pp. 1–22. https://journals.lub.lu.se/pa/article/view/19024.

Blythman, Joanna. 2018. 'The Quorn Revolution: The Rise of Ultra-Processed Fake Meat'. The Guardian. https://www.theguardian.com/lifeandstyle/2018/feb/12/quorn-revolution-rise-ultra-processed-fake-meat.

Bobier, Christopher. 2019. 'Should Moral Vegetarians Avoid Eating Vegetables?' *Food Ethics* 5 (1): pp. 1–15. doi: 10.1007/s41055-019-00062-4.

Bohrer, Benjamin M. 2019. 'An Investigation of the Formulation and Nutritional Composition of Modern Meat Analogue Products'. *Food Science and Human Wellness* 8 (4): pp. 320–329. doi: 10.1016/j.fshw.2019.11.006.

Boppré, Michael, and Richard I. Vane-Wright. 2019. 'Welfare Dilemmas Created by Keeping Insects in Captivity'. In *The Welfare of Invertebrate Animals*, edited by Claudio Carere and Jennifer Mather, pp. 23–68. Dordrecht: Springer. doi: 10.1007/978-3-030-13947-6_3.

Bovenkerk, Bernice, Eva Meijer, and Hanneke Nijland. 2020. 'Veganisme of Menselijk Diervoer? Een Niet-Antropocentrische Benadering van het Wereldvoedselprobleem. In *Tien Miljard Monden: Hoe We de Wereld Gaan Voeden in 2050*, edited by Ingrid de Zwarte and Jeroen Candel, pp. 346–352. Amsterdam: Prometheus.

Bradshaw, Karen. 2020. *Wildlife as Property Owners: A New Conception of Animal Rights*. Chicago: University of Chicago Press.

Broad, Garrett M. 2019. 'Plant-Based and Cell-Based Animal Product Alternatives: An Assessment and Agenda for Food Tech Justice'. *Geoforum* 107: pp. 223–226. doi: 10.1016/j.geoforum.2019.06.014.

Bruckner, Donald W. 2015. 'Strict Vegetarianism is Immoral'. In *The Moral Complexities of Eating Meat*, edited by Ben Bramble and Bob Fischer, pp. 30–47. Oxford: Oxford University Press. doi: 10.1093/acprof:oso/9780199353903.003.0003.

Budolfson, Mark. 2018. 'Food, the Environment, and Global Justice'. In *The Oxford Handbook of Food Ethics*, edited by Anne Barnhill, Mark Budolfson, and Tyler Doggett, pp. 67–94. Oxford: Oxford University Press. doi: 10.1093/oxfordhb/9780199372263.013.4.

Calhoone, Lawrence. 2009. 'Hunting as a Moral Good'. *Environmental Values* 18 (1): pp. 67-89. doi: 10.3197/096327109X404771.
Callicott, J. Baird. 2015. 'The Environmental Omnivore's Dilemma'. In *The Moral Complexities of Eating Meat*, edited by Ben Bramble and Bob Fischer, pp. 48-64. Oxford: Oxford University Press. doi: 10.1093/acprof:oso/9780199353903.001.0001.
Carder, Gemma. 2017. 'A Preliminary Investigation into the Welfare of Lobsters in the UK'. *Animal Sentience* 16 (19): pp. 1-9. doi: 10.51291/2377-7478.1262.
Carere, Claudio, and Jennifer Mather. 2019. 'Why Invertebrate Welfare?' In *The Welfare of Invertebrate Animals*, edited by Claudio Carere and Jennifer Mather, pp. 1-6. Dordrecht: Springer. doi: 10.1007/978-3-030-13947-6_1.
Ceurstemont, Sandrine. 2017. 'Make Your Own Meat with Open-Source Cells—No Animals Necessary'. New Scientist. https://www.newscientist.com/article/mg23331080-700-make-your-own-meat-with-open-source-cells-no-animals-necessary/.
Chiles, Robert M., Garrett Broad, Mark Gagnon, Nicole Negowetti, Leland Glenna, Megan A. M. Griffin, Lina Tami-Barrera, Siena Baker, and Kelly Becky. 2021. 'Democratizing Ownership and Participation in the 4th Industrial Revolution: Challenges and Opportunities in Cellular Agriculture'. *Agriculture and Human Values* 38: pp. 943-961. doi: 10.1007/s10460-021-10237-7.
Ciocchetti, Christopher. 2012. 'Veganism and Living Well'. *Journal of Agricultural and Environmental Ethics* 25 (3): pp. 405-417. doi: 10.1007/s10806-011-9307-5.
Clark, Samuel. 2015. 'Good Work'. *Journal of Applied Philosophy* 34 (1): pp. 61-73. doi: 10.1111/japp.12137.
Cochrane, Alasdair. 2012. *Animal Rights Without Liberation: Applied Ethics and Human Obligations*. New York: Columbia University Press.
Cochrane, Alasdair. 2013. 'From Human Rights to Sentient Rights'. *Critical Review of International Social and Political Philosophy* 16 (5): pp. 655-675. doi: 10.1080/13698230.2012.691235.
Cochrane, Alasdair. 2016. 'Labour Rights for Animals'. In *The Political Turn in Animal Ethics*, edited by Robert Garner and Siobhan O'Sullivan, pp. 15-32. London: Rowman & Littlefield International.
Cochrane, Alasdair. 2018. *Sentientist Politics: A Theory of Global Inter-Species Justice*. Oxford: Oxford University Press.
Cochrane, Alasdair. 2020. 'Good Work for Animals'. In *Animal Labour: A New Frontier of Interspecies Justice?*, edited by Charlotte E. Blattner, Kendra Coulter, and Will Kymlicka, pp. 48-64. Oxford: Oxford University Press. doi: 10.1093/oso/9780198846192.001.0001.
Cochrane, Alasdair, Robert Garner, and Siobhan O'Sullivan. 2018. 'Animal Ethics and the Political'. *Critical Review of International Social and Political Philosophy* 21 (2): pp. 261-277. doi: 10.1080/13698230.2016.1194583.
Cole, Matthew, and Karen Morgan. 2013. 'Engineering Freedom? A Critique of Biotechnological Routes to Animal Liberation'. *Configurations* 21 (2): pp. 201-229. doi: 10.1353/con.2013.0015.
Comstock, Gary. 2012. 'Ethics and Genetically Modified Food'. In *The Philosophy of Food*, edited by David Kaplan, pp. 109-124. Berkeley: University of California Press. doi: 10.1525/9780520951976-008.
Cooke, Steve. 2017a. 'Animal Kingdoms: On Habitat Rights for Wild Animals'. *Environmental Values* 26 (1): pp. 53-72. doi: 10.3197/096327117x14809634978555.
Cooke, Steve. 2017b. 'Imagined Utopias: Animal Rights and the Moral Imagination'. *The Journal of Political Philosophy* 25 (4): pp. e1-e18. doi: 10.1111/jopp.12136.
Coulter, Kendra. 2016. *Animals, Work, and the Promise of Interspecies Solidarity*. Basingstoke: Palgrave Macmillan.
Coulter, Kendra. 2017. 'Humane Jobs: A Political Economic Vision for Interspecies Solidarity and Human-Animal Wellbeing'. *Politics and Animals* 3: pp. 31-41. https://journals.lub.lu.se/pa/article/view/16589.
Coulter, Kendra. 2020. 'Toward Humane Jobs and Work-Lives for Animals'. In *Animal Labour: A New Frontier of Interspecies Justice?*, edited by Charlotte E. Blattner, Kendra Coulter, and Will Kymlicka, pp. 29-47. Oxford: Oxford University Press. doi: 10.1093/oso/9780198846192.003.0002.
Coulter, Kendra, and Josh Milburn. 2022. '(Not) Serving Animals and Aiming Higher: Cultivating Ethical Plant-Based Businesses and Humane Jobs'. In *Animals and Business Ethics*, edited by Natalie Thomas, pp. 43-66. Basingstoke: Palgrave Macmillan. doi: 10.1007/978-3-030-97142-7_3.

Davis, Steven L. 2003. 'The Least Harm Principle May Require that Humans Consume a Diet Containing Large Herbivores, Not a Vegan Diet'. *Journal of Agricultural and Environmental Ethics* 16 (4): pp. 387–394. doi: 10.1023/a:1025638030686.
Deckers, Jan. 2016. *Animal (De)liberation: Should the Consumption of Animal Products Be Banned?* London: Ubiquity Press.
Deckha, Maneesha. 2020. 'Veganism, Dairy, and Decolonization'. *Journal of Human Rights and the Environment* 11 (2): pp. 244–267. doi: 10.4337/jhre.2020.02.05.
Delon, Nicolas. 2020. 'The Meaning of Animal Labour'. In *Animal Labour: A New Frontier of Interspecies Justice?*, edited by Charlotte E. Blattner, Kendra Coulter, and Will Kymlicka, pp. 160–180. Oxford: Oxford University Press. doi: 10.1093/oso/9780198846192.003.0008.
Derham, Tristan, and Freya Mathews. 2020. 'Elephants as Refugees'. *People and Nature* 2 (1): pp. 103–110. doi: 10.1002/pan3.10070.
Devlin, Patrick. 1965. *The Enforcement of Morals*. Oxford: Oxford University Press.
Dickstein, Jonathan, and Jan Dutkiewicz. 2021. 'The Ism in Veganism: The Case for a Minimal Practice-Based Definition'. *Food Ethics* 6 (2): pp. 1–19. doi: 10.1007/s41055-020-00081-6.
Diggles, B. K. 2019. 'Review of Some Scientific Issues Related to Crustacean Welfare'. *ICES Journal of Marine Science* 76 (1): pp. 66–81. doi: 10.1093/icesjms/fsy058.
Doggett, Tyler. 2018. 'Moral Vegetarianism'. In *The Stanford Encyclopedia of Philosophy*, edited by Edward N. Zalta. https://plato.stanford.edu/entries/vegetarianism/.
Donaldson, Sue, and Will Kymlicka. 2011. *Zoopolis: A Political Theory of Animal Rights*. Oxford: Oxford University Press.
Donaldson, Sue, and Will Kymlicka. 2015. 'Farmed Animal Sanctuaries: The Heart of the Movement?' *Politics and Animals* 1: pp. 50–74. https://journals.lub.lu.se/pa/article/view/15045.
Donaldson, Sue, and Will Kymlicka. 2017. 'Animals in Political Theory'. In *The Oxford Handbook of Animal Studies*, edited by Linda Kalof, pp. 43–64. Oxford: Oxford University Press. doi: 10.1093/oxfordhb/9780199927142.013.33.
Donaldson, Sue, and Will Kymlicka. 2020. 'Animal Labour in a Post-Work Society'. In *Animal Labour: A New Frontier of Interspecies Justice?*, edited by Charlotte E. Blattner, Kendra Coulter, and Will Kymlicka, pp. 207–228. Oxford: Oxford University Press. doi: 10.1093/oso/9780198846192.001.0001.
Dunayer, Joan. 1995. 'Sexist Words, Speciesist Roots'. In *Animals and Women: Feminist Theoretical Explorations*, edited by Carol J. Adams and Josephine Donovan, pp. 11–31. Durham, NC: Duke University Press. doi: 10.1215/9780822381952-002.
Dutkiewicz, Jan. 2019. 'Socialize Lab Meat'. *Jacobin*. https://jacobinmag.com/2019/08/lab-meat-socialism-green-new-deal.
Dutkiewicz, Jan, and Elan Abrell. 2021. 'Sanctuary to Table Dining: Cellular Agriculture and the Ethics of Cell Donor Animals'. *Politics and Animals* 7: pp. 1–15. https://journals.lub.lu.se/pa/article/view/22252.
Dworkin, Ronald. 1984. 'Rights as trumps'. In *Theories of Rights*, edited by Jeremy Waldron, pp. 153–167. Oxford: Oxford University Press.
Elson, H. A. 2015. 'Poultry Welfare in Intensive and Extensive Production Systems'. *World's Poultry Science Journal* 71 (3): pp. 449–460. doi: 10.1017/S0043933915002172.
Elwood, Robert W. 2011. 'Pain and Suffering in Invertebrates?' *ILAR Journal* 52 (2): pp. 175–184. doi: 10.1093/ilar.52.2.175.
Elwood, Robert W. 2019. 'Assessing the Potential for Pain in Crustaceans and Other Invertebrates'. In *The Welfare of Invertebrate Animals*, edited by Claudio Carere and Jennifer Mather, pp. 147–178. Dordrecht: Springer.
Emmerman, Karen S. 2019. 'What's Love Got to Do with It? An Ecofeminist Approach to Inter-Animal and Intra-Cultural Conflicts of Interest'. *Ethical Theory and Moral Practice* 22 (1): pp. 77–91. doi: 10.1007/s10677-019-09978-6.
Evans, Matthew. 2019. *On Eating Meat: The Truth About Its Production and the Ethics of Eating It*. Millers Point: Murdoch Books.
Fabre, Cecile. 2022. *Spying Through a Glass Darkly: The Ethics of Espionage and Counter-Intelligence*. Oxford: Oxford University Press.
Fairlie, Simon. 2010. *Meat: A Benign Extravagance*. East Meon, Hampshire: Permanent Publications.
FAO. 2019. 'The State of Food Security and Nutrition in the World 2019' (report). http://www.fao.org/3/ca5162en/ca5162en.pdf.

Firth, Henry, and Ian Theasby. 2019. *Bish Bash BOSH!* London: HQ.
Fischer, Bob. 2016a. 'Bugging the Strict Vegan'. *Journal of Agricultural and Environmental Ethics* 29 (2): pp. 255–263. doi: 10.1007/s10806-015-9599-y.
Fischer, Bob. 2016b. 'What if Klein & Barron Are Right About Insect Sentience?' *Animal Sentience* 9 (8): pp. 1–6. doi: 10.51291/2377-7478.1138.
Fischer, Bob. 2020. *The Ethics of Eating Animals: Usually Bad, Sometimes Wrong, Often Permissible.* Abingdon: Routledge.
Fischer, Bob, and Andy Lamey. 2018. 'Field Deaths in Plant Agriculture'. *Journal of Agricultural and Environmental Ethics* 31 (4): pp. 409–428. doi: 10.1007/s10806-018-9733-8.
Fischer, Bob, and Josh Milburn. 2019. 'In Defence of Backyard Chickens'. *Journal of Applied Philosophy* 36 (1): pp. 108–123. doi: 10.1111/japp.12291.
Fischer, Bob, and Burkay Ozturk. 2017. 'Facsimiles of Flesh'. *Journal of Applied Philosophy* 34 (4): pp. 489–497. doi: 10.1111/japp.12223.
Fitzgerald, Amy J., Linda Kalof, and Thomas Dietz. 2009. 'Slaughterhouses and Increased Crime Rates: An Empirical Analysis of the Spillover from "The Jungle" into the Surrounding Community'. *Organization & Environment* 22 (2): pp. 158–184. doi: 10.1177/1086026609338164.
Francione, Gary L. 1996. *Rain Without Thunder: The Ideology of the Animal Rights Movement.* Philadelphia, PA: Temple University Press.
Francione, Gary L. 2000. *Introduction to Animal Rights: Your Child or the Dog.* Philadelphia, PA: Temple University Press.
Francione, Gary L. 2009. *Animals as Persons: Essays on the Abolition of Animal Exploitation.* New York: Columbia University Press.
Francione, Gary L. 2012. 'Animal Welfare, Happy Meat, and Veganism as the Moral Baseline'. In *The Philosophy of Food*, edited by David Kaplan, pp. 169–179. Berkeley: University of California Press. doi: 10.1525/9780520951976-011.
Francione, Gary L., and Anna Charlton. 2015. *Animal Rights: The Abolitionist Approach.* Louisville: Exempla Press.
Francione, Gary L., and Robert Garner. 2010. *The Animal Rights Debate: Abolition or Regulation?* New York: Columbia University Press.
Fredrickson, T. N. 1987. 'Ovarian Tumors of the Hen'. *Environmental Health Perspectives* 73: pp. 35–51. doi: 10.1289/ehp.877335.
Friedrich, Bruce. 2020. 'Foreword'. In *Moo's Law: An Investor's Guide to the New Agrarian Revolution*, by Jim Mellon, pp. 20–23. Sudbury: Fruitful Publications.
Gaard, Greta. 2013. 'Toward a Feminist Postcolonial Milk Studies'. *American Quarterly* 65 (3): pp. 595–618. doi: 10.1353/aq.2013.0040.
Galli, Roberta, Grit Preusse, Christian Schnabel, Thomas Bartels, Kerstin Cramer, Maria-Elisabeth Krautwald-Junghanns, Edmund Koch, and Gerald Steiner. 2018. 'Sexing of Chicken Eggs by Fluorescence and Raman Spectroscopy through the Shell Membrane'. *PLOS One* 13 (2): p. e0192554. doi: 10.1371/journal.pone.0192554.
Garner, Robert. 2012. 'Toward a Theory of Justice for Animals'. *Journal of Animal Ethics* 2 (1): pp. 98–104. doi: 10.5406/janimalethics.2.1.0098.
Garner, Robert. 2013. *A Theory of Justice for Animals: Animal Rights in a Nonideal World.* Oxford: Oxford University Press.
Garner, Robert. 2015. 'Welfare, rights, and non-ideal theory'. In *The Ethics of Killing Animals*, edited by Tatjana Višak and Robert Garner, pp. 215–228. Oxford: Oxford University Press. doi: 10.1093/acprof:oso/9780199396078.001.0001.
Garrido, Claudia, and Antonio Nanetti. 2019. 'Welfare of Managed Honey Bees'. In *The Welfare of Invertebrate Animals*, edited by Claudio Carere and Jennifer Mather, pp. 69–104. Dordrecht: Springer.
George, Kathryn Paxton. 1994. 'Should Feminists be Vegetarians?' *Signs* 19 (2): pp. 405–434. doi: 10.1086/494889.
George, Kathryn Paxton. 2000. *Animal, Vegetable, or Woman? A Feminist Critique of Ethical Vegetarianism.* New York: SUNY Press.
Gottlieb, Robert, and Anupama Joshi. 2010. *Food Justice.* Cambridge, MA: The MIT Press.

Guerrero, Alex. 2007. 'Don't Know, Don't Kill: Moral Ignorance, Culpability, and Caution'. *Philosophical Studies* 136 (1): pp. 59–97. doi: 10.1007/s11098-007-9143-7.
Hadley, John. 2015. *Animal Property Rights: A Theory of Habitat Rights for Wild Animals*. Lanham, MD: Lexington Books.
Hadley, John. 2019a. *Animal Neopragmatism: From Welfare to Rights*. Basingstoke: Palgrave Macmillan.
Hadley, John. 2019b. 'Does a Painless Death Harm an Invertebrate?' *Australian Zoologist* 40 (1): pp. 151–157. doi: 10.7882/AZ.2017.038.
Hayek, Matthew, and Jan Dutkiewicz. 2021. 'Yes, Plant-Based Meat Is Better for the Planet'. Vox. https://www.vox.com/22787178/beyond-impossible-plant-based-vegetarian-meat-climate-environmental-impact-sustainability.
Heller, Martin C., and Gregory A. Keoleian. 2018. 'Beyond Meat's Beyond Burger Life Cycle Assessment: A Detailed Comparison Between a Plant-Based and an Animal-Based Protein Source'. CSS Report: CSS18-10. https://css.umich.edu/publications/research-publications/beyond-meats-beyond-burger-life-cycle-assessment-detailed.
Holdier, A. G. 2016. 'The Pig's Squeak: Towards a Renewed Aesthetic Argument for Veganism'. *Journal of Agricultural and Environmental Ethics* 29 (4): pp. 631–642. doi: 10.1007/s10806-016-9624-9.
Horvath, Kelsey, Dario Angeletti, Guiseppe Nascetti, and Claudio Carere. 2013. 'Invertebrate Welfare: An Overlooked Issue'. *Annali dell'Istituto Superiore di Sanità* 49 (1): pp. 9–17. doi: 10.4415/ANN_13_01_04.
Jacquet, Jennifer, Jeff Sebo, and Max Elder. 2017. 'Seafood in the Future: Bivalves are Better'. *Solutions* 8 (1): pp. 27–32. https://thesolutionsjournal.com/2017/01/11/seafood-future-bivalves-better/.
Johnson, P. A., C. S. Stephens, and J. R. Giles. 2015. 'The Domestic Chicken: Causes and Consequences of an Egg a Day'. *Poultry Science* 94 (4): pp. 816–820. doi: 10.3382/ps/peu083.
Johnson, Patricia A., and James R. Giles. 2013. 'The Hen as a Model of Ovarian Cancer'. *Nature Reviews Cancer* 13 (6): pp. 432–436. doi: 10.1038/nrc3535.
Jones, Robert C. 2017. 'The Precautionary Principle: A Cautionary Note'. *Animal Sentience* 16 (15): pp. 1–3. doi: 10.51291/2377-7478.1250.
Joy, Melanie. 2009. *Why We Love Dogs, Eat Pigs, and Wear Cows: An Introduction to Carnism*. San Francisco: Conari.
Katz-Rosene, Ryan M., and Sarah J. Martin, eds. 2020. *Green Meat? Sustaining Eaters, Animals, and the Planet*. Kingston and Montreal: McGill-Queen's University Press.
Kazez, Jean. 2018. 'The Taste Question in Animal Ethics'. *Journal of Applied Philosophy* 35 (4): pp. 661–674. doi: 10.1111/japp.12278.
Kianpour, Connor Kayhan. 2020. 'Cetacean Property: A Hegelish Account of Nonhuman Property'. *Politics and Animals* 6: pp. 23–36. https://journals.lub.lu.se/pa/article/view/19664.
Kim, Claire Jean. 2015. *Dangerous Crossings: Race, Species, and Nature in a Multicultural Age*. Cambridge: Cambridge University Press.
Klein, Colin. 2017. 'Precaution, Proportionality and Proper Commitments'. *Animal Sentience* 16 (9): pp. 1–3. doi: 10.51291/2377-7478.1232.
Klein, Colin, and Andrew B. Barron. 2016. 'Insects Have the Capacity for Subjective Experience'. *Animal Sentience* 9 (1): pp. 1–52. doi: 10.51291/2377-7478.111.
Knight, Andrew, Claire Parkinson, Patricia MacCormack, Richard Twine, Mariah Peixoto, Helena Pedersen, Jonna Håkansson, Dorna Behdadi, Thomas Laurien, Björn Olsen, Lina Gustafsson, Kerstin Malm, Elin Pöllänen, Tânia Regina Vizachri, Thiago Pires-Oliveira. 2021. 'Transition to Plant-Based Diets Will Help Us Fight Pandemics'. *Revista Latino-Americana de Direitos da Natureza e dos Animais* 3 (2): pp. 143–148. https://periodicos.ucsal.br/index.php/rladna/article/view/845/0.
Kravitz, Melissa. 2018. 'Are Scallops Actually Vegan?' Vice. https://www.vice.com/en/article/qvxznq/are-scallops-vegan.
Kymlicka, Will. 1989. 'Liberal Individualism and Liberal Neutrality'. *Ethics* 99 (4): pp. 883–905. doi: 10.1086/293125.
Kymlicka, Will. 2017. 'Social Membership: Animal Law Beyond the Property/Personhood Impasse'. *Dalhousie Law Journal* 40 (1): pp. 123–155. doi: https://digitalcommons.schulichlaw.dal.ca/dlj/vol40/iss1/4/.

Lagerlund, Henrik. 2018. 'Food Ethics in the Middle Ages'. In *The Oxford Handbook of Food Ethics*, edited by Anne Barnhill, Mark Budolfson, and Tyler Doggett, pp. 759–772. Oxford: Oxford University Press. doi: 10.1093/oxfordhb/9780199372263.013.16.
Lamey, Andy. 2016. 'Subjective Experience and Moral Standing'. *Animal Sentience* 9 (7): pp. 1–3. doi: 10.51291/2377-7478.1136.
Lamey, Andy. 2019. *Duty and the Beast: Should We Eat Meat in the Name of Animal Rights?* Cambridge: Cambridge University Press.
Lance, Stephanie. 2020. 'The Cost of Production: Animal Welfare and the Post-Industrial Slaughterhouse in Margaret Atwood's *Oryx and Crake*'. *MOSF Journal of Science Fiction* 4 (1): pp. 60–74. https://publish.lib.umd.edu/?journal=scifi&page=article&op=view&path%5B%5D=518.
Leopold, Aldo. 1949. *A Sand County Almanac: And Sketches Here and There*. Oxford: Oxford University Press.
Lestar, Tamas. 2021. 'Why Imported Veg Is Still More Sustainable than Local Meat'. The Conversation. https://theconversation.com/why-imported-veg-is-still-more-sustainable-than-local-meat-159943.
List, Charles. 2018. 'The New Hunter and Local Food'. In *The Oxford Handbook of Food Ethics*, edited by Anne Barnhill, Mark Budolfson, and Tyler Doggett, pp. 170–188. Oxford: Oxford University Press. doi: 10.1093/oxfordhb/9780199372263.013.18.
Lomasky, Loren. 2013. 'Is It Wrong to Eat Animals?' *Social Philosophy and Policy* 30 (1–2): pp. 177–200. doi: 10.1017/s0265052513000083.
Machan, Tibor R. 2004. *Putting Humans First: Why We Are Nature's Favorite*. Lanham: Rowman & Littlefield.
McMahan, Jeff. 2002. *The Ethics of Killing: Problems at the Margins of Life*. Oxford: Oxford University Press.
McMahan, Jeff. 2008. 'Eating Animals the Nice Way'. *Daedalus* 137 (1): pp. 66–76. doi: 10.1162/daed.2008.137.1.66.
Mares, Teresa M., and Devon G. Peña. 2011. 'Environmental and Food Justice: Toward Local, Slow, and Deep Food Systems'. In *Cultivating Food Justice: Race, Class, and Sustainability*, edited by Alison Hope Alkon and Julian Agyeman, pp. 197–220. Cambridge, MA: The MIT Press. doi: 10.7551/mitpress/8922.003.0014.
Marks, Joel. 2017. 'Changing the Subject'. *Animal Sentience* 16 (5): pp. 1–6. doi: 10.51291/2377-7478.1221.
Martin, Angela K. 2022. 'Animal Research that Respects Animal Rights: Extending Requirements for Research with Humans to Animals'. *Cambridge Quarterly of Healthcare Ethics* 31 (1): pp. 59–72. doi: 10.1017/S0963180121000499.
Matheny, Gaverick. 2003. 'Least Harm: A Defense of Vegetarianism from Steven Davis's Omnivorous Proposal'. *Journal of Agricultural and Environmental Ethics* 16 (5): pp. 505–511. doi: 10.1023/A:1026354906892.
Meijer, Eva. 2019. *When Animals Speak: Toward an Interspecies Democracy*. New York: NYU Press.
Melina, Vesanto, Winston Craig, and Susan Levin. 2016. 'Position of the Academy of Nutrition and Dietetics: Vegetarian Diets'. *Journal of the Academy of Nutrition and Dietetics* 116 (1): pp. 1970–1980. doi: 10.1016/j.jand.2016.09.025.
Mellon, Jim. 2020. *Moo's Law: An Investor's Guide to the New Agrarian Revolution*. Sudbury: Fruitful Publications.
Melzener, Lea, Karin E. Verzijden, A. Jasmin Buijs, Mark J. Post, Joshua E Flack. 2021. 'Cultured Beef: From Small Biopsy to Substantial Quantity'. *Journal of the Science of Food and Agriculture* 101 (1): pp. 7–14. doi: 10.1002/jsfa.10663.
Meyers, C. D. 2013. 'Why It Is Morally Good to Eat (Certain Kinds of) Meat: The Case for Entomophagy'. *Southwest Philosophy Review* 29 (1): pp. 119–126. doi: 10.5840/swphilreview201329113.
Milburn, Josh. 2015. 'Not Only Humans Eat Meat: Companions, Sentience, and Vegan Politics'. *Journal of Social Philosophy* 46 (4): pp. 449–462. doi: 10.1111/josp.12131.
Milburn, Josh. 2016. 'Chewing Over *In Vitro* Meat: Animal Ethics, Cannibalism and Social Progress'. *Res Publica* 22 (3): pp. 249–265. doi: 10.1007/s11158-016-9331-4.
Milburn, Josh. 2017a. 'Nonhuman Animals as Property Holders: An Exploration of the Lockean Labour-Mixing Account'. *Environmental Values* 26 (5): pp. 629–648. doi: 10.3197/096327117x15002190708155.

Milburn, Josh. 2017b. 'Robert Nozick on Nonhuman Animals: Rights, Value and the Meaning of Life'. In *Ethical and Political Approaches to Nonhuman Animal Issues*, edited by Andrew Woodhall and Gabriel Garmendia da Trindade, pp. 97–120. Basingstoke: Palgrave Macmillan. doi: 10.1007/978-3-319-54549-3_5.

Milburn, Josh. 2018. 'Death-Free Dairy? The Ethics of Clean Milk'. *Journal of Agricultural and Environmental Ethics* 31 (2): pp. 261–279. doi: 10.1007/s10806-018-9723-x.

Milburn, Josh. 2019. 'Review: *When Animals Speak*'. *Metapsychology* 23 (51). https://metapsychology.net/index.php/book-review/when-animals-speak/.

Milburn, Josh. 2020. 'A Novel Argument for Vegetarianism? Zoopolitics and Respect for Animal Corpses'. *Animal Studies Journal* 9 (2): pp. 240–259. doi: 10.14453/asj/v9.i2.10.

Milburn, Josh. 2022a. 'Ethics of Meat Replacements'. In *Meat and Meat Replacements: An Interdisciplinary Assessment of Current Status and Future Directions*, edited by Herbert L. Meiselman and Jose M. Lorenzo. pp. 257–280 Amsterdam: Elsevier.

Milburn, Josh. 2022b. *Just Fodder: The Ethics of Feeding Animals*. Montreal and Kingston: McGill-Queen's University Press.

Milburn, Josh. 2022c. 'Should Vegans Compromise?' *Critical Review of International Social and Political Philosophy* 25 (2): pp. 281–293. doi: 10.1080/13698230.2020.1737677.

Milburn, Josh. Forthcoming. 'Freeganism: A (Cautious) Defense'. In *New Omnivorism and Strict Veganism: Critical Perspectives*, edited by Cheryl Abbate and Christopher Bobier. Abingdon: Routledge.

Milburn, Josh, and Christopher Bobier. 2022. 'New Omnivorism: A Novel Approach to Food and Animal Ethics'. *Food Ethics* 7 (5): pp. 1–17. doi: 10.1007/s41055-022-00098-z.

Milburn, Josh, and Alasdair Cochrane. 2021. 'Should We Protect Animals from Hate Speech?' *Oxford Journal of Legal Studies* 41 (4): pp. 1149–1172. doi: 10.1093/ojls/gqab013.

Milburn, Josh, and Bob Fischer. 2021. 'The Freegan Challenge to Veganism'. *Journal of Agricultural and Environmental Ethics* 34 (17): pp. 1–19. doi: 10.1007/s10806-021-09859-y.

Mill, John Stuart. 2008. 'On Liberty'. In *On Liberty and Other Essays*, edited by John Gray, pp. 1–128. Oxford: Oxford University Press.

Miller, John. 2012. 'In Vitro Meat: Power, Authenticity, and Vegetarianism'. *Journal for Critical Animal Studies* 10 (4): pp. 41–63. http://www.criticalanimalstudies.org/wp-content/uploads/2012/12/JCAS-Volume-10-Issue-4-2012.pdf.

Miller, John. Forthcoming. *Utopian Protein*. Ms.

Milligan, Tony. 2015. 'The Political Turn in Animal Rights'. *Politics and Animals* 1: pp. 6–15. https://journals.lub.lu.se/pa/article/view/13512.

Monteiro, Carlos Augusto, Geoffrey Cannon, Jean-Claude Moubarac, Renata Bertazzi Levy, Maria Laura C. Louzada, and Patrícia Constante Jaime. 2017. 'The UN Decade of Nutrition, the NOVA Food Classification and the Trouble with Ultra-Processing'. *Public Health Nutrition* 21 (1): pp. 5–17. doi: 10.1017/S1368980017000234.

Mosquera, Julia. 2016. 'Are Nonhuman Animals Owed Compensation for the Wrongs Committed to Them?' In *Intervention or Protest: Acting for Nonhuman Animals*, edited by Andrew Woodhall and Gabriel Garmendia da Trindade, pp. 213–242. Delaware: Vernon Press.

Mufson, Beckett. 2018. 'This Guy Served His Friends Tacos Made from His Own Amputated Leg'. Vice. https://www.vice.com/en/article/gykmn7/legal-ethical-cannibalism-human-meat-tacos-reddit-wtf.

Nagel, Thomas. 1974. 'What Is It Like to Be a Bat?' *The Philosophical Review* 83 (4): pp. 435–450. doi: 10.2307/2183914.

Newman, Lenore. 2020. 'The Promise and Peril of "Cultured Meat"'. In *Green Meat? Sustaining Eaters, Animals, and the Planet*, edited by Ryan M. Katz-Rosene and Sarah J. Martin, pp. 169–184. Kingston and Montreal: McGill-Queen's University Press.

Nguyen, Angela, and Michael J. Platow. 2021. '"I'll Eat Meat Because That's What We Do": The Role of National Norms and National Social Identification on Meat Eating'. *Appetite* 164 (105287): pp. 1–10. doi: 10.1016/j.appet.2021.105287.

Nozick, Robert. 1974. *Anarchy, State and Utopia*. New York: Basic Books.

Nozick, Robert. 1981. *Philosophical Explanations*. Cambridge, MA: Harvard University Press.

Nussbaum, Martha. 2007. *Frontiers of Justice: Disability, Nationality, Species Membership*. Cambridge, MA: Harvard University Press.
O'Connor, Anahad. 2019. 'Fake Meat Vs. Real Meat'. The New York Times. https://www.nytimes.com/2019/12/03/well/eat/fake-meat-vs-real-meat.html.
O'Sullivan, Siobhan. 2011. *Animals, Equality, and Democracy*. Basingstoke: Palgrave Macmillan.
OECD. 2016. 'Antimicrobial Resistance: Policy Insights' (report). https://www.oecd.org/health/health-systems/AMR-Policy-Insights-November2016.pdf.
Palmer, Clare. 2010. *Animal Ethics in Context*. New York: Columbia University Press.
Pérez, José Luis Rey. 2018. *Los Derechos de los Animales en Serio*. Madrid: Dykinson.
Pérez, José Luis Rey. 2019. 'Labour Rights for Animals?' Presentation given at the 2019 MANCEPT Workshops.
Plato. 1992. *Republic*, translated by G. M. A. Grube, revised by C. D. C. Reeve. Cambridge: Hackett.
Plumwood, Val. 2000. 'Integrating Ethical Frameworks for Animals, Humans, and Nature: A Critical Feminist Eco-Socialist Analysis'. *Ethics and the Environment* 5 (2): pp. 285–322. doi: 10.1016/s1085-6633(00)00033-4.
Poinski, Megan. 2022. 'Consumers Want Animal-Free Dairy Because It's Better for Cows, Study Finds'. *Food Dive*. https://www.fooddive.com/news/consumers-want-animal-free-dairy-because-its-better-for-cows-study-finds/618675/.
Pollo, Simone, and Augusto Vitale. 2019. 'Invertebrates and Humans: Science, Ethics, and Policy'. In *The Welfare of Invertebrate Animals*, edited by Claudio Carere and Jennifer Mather, pp. 7–22. Dordrecht: Springer. doi: 10.1007/978-3-030-13947-6_2.
Poore, Joseph, and Thomas Nemecek. 2018. 'Reducing Food's Environmental Impacts Through Producers and Consumers'. *Science* 360 (6392): pp. 987–992. doi: 10.1126/science.aaq0216.
Porcher, Jocelyne. 2017. *The Ethics of Animal Labor: A Collaborative Utopia*. Basingstoke: Palgrave Macmillan.
Potts, Annie. 2016. 'What Is Meat Culture?' In *Meat Culture*, edited by Annie Potts, pp. 1–30. Leiden: Brill. doi: 10.1163/9789004325852_002.
Powers, Madison, Ruth Faden, and Yashar Saghai. 2012. 'Liberty, Mill, and the Framework of Public Health Ethics'. *Public Health Ethics* 5 (1): pp. 6–15. doi: 10.1093/phe/phs002.
Rawls, John. 1999. *A Theory of Justice* (revised edition). Cambridge, MA: Harvard University Press.
Regan, Tom. 1983. *The Case for Animal Rights*. Berkeley: University of California Press.
Regan, Tom. 2004a. *Empty Cages: Facing the Challenge of Animal Rights*. Lanham: Rowman & Littlefield.
Regan, Tom. 2004b. *The Case for Animal Rights* (updated edition). Berkeley: University of California Press.
Rollin, Bernard E. 2011. 'Animal Pain: What It Is and Why It Matters'. *The Journal of Ethics* 15 (4): pp. 425–437. doi: 10.1007/s10892-010-9090-y.
Romanov, M. N. 2001. 'Genetics of Broodiness in Poultry—A Review'. *Asian-Australian Journal of Animal Sciences* 14 (11): pp. 1647–1654. doi: 10.5713/ajas.2001.1647.
Rousseau, Jean-Jacques. 1968. *The Social Contract*, translated by Maurice Cranston. London: Penguin.
Rust, Niki A., Lucia Rehackova, Francis Naab, Amber Abrams, Courtney Hughes, Bethann Garramon Merkle, Beth Clark, and Sophie Tindale. 2021. 'What Does the UK Public Want Farmland to Look Like?' *Land Use Policy* 106 (105445): pp. 1–14. doi: 10.1016/j.landusepol.2021.105445.
Sachs, Benjamin. 2019. 'Teleological Contractarianism'. *Journal of Social Philosophy* 50 (1): pp. 91–112. doi: 10.1111/josp.12266.
Sandler, Ronald. 2015. *Food Ethics: The Basics*. Abingdon: Routledge.
Santo, Raychel E., Brent F. Kim, Sarah E. Goldman, Jan Dutkiewicz, Erin M. B. Biehl, Martin W. Bloem, Roni A. Neff, and Keeve E. Nachman. 2020. 'Considering Plant-Based Meat Substitutes and Cell-Based Meats: A Public Health and Food Systems Perspective'. *Frontiers in Sustainable Food Systems* 4 (134): pp. 1–23. doi: 10.3389/fsufs.2020.00134.
Scarborough, Peter, Paul N. Appleby, Anja Mizdrak, Adam D. M. Briggs, Ruth C. Travis, Kathryn E. Bradbury, and Timothy J. Key. 2014. 'Dietary Greenhouse Gas Emissions of Meat-Eaters, Fish-Eaters, Vegetarians and Vegans in the UK'. *Climatic Change* 125 (2): pp. 179–192. doi: 10.1007/s10584-014-1169-1.
Schaefer, G. Owen, and Julian Savulescu. 2014. 'The Ethics of Producing *In Vitro* Meat'. *Journal of Applied Philosophy* 31 (2): pp. 188–202. doi: 10.1111/japp.12056.

Schmitz, Friederike. 2016. 'Animal Ethics and Human Institutions: Integrating Animals into Political Theory'. In *The Political Turn in Animal Ethics*, edited by Robert Garner and Siobhan O'Sullivan, pp. 33–49. London: Rowman & Littlefield International.

Scotton, Guy. 2017. 'Interspecies Atrocities and the Politics of Memory'. In *Ethical and Political Approaches to Nonhuman Animal Issues*, edited by Andrew Woodhall and Gabriel Garmendia da Trindade, pp. 305–326. Basingstoke: Palgrave Macmillan. doi: 10.1007/978-3-319-54549-3_13.

Sebo, Jeff. 2018. 'The Moral Problem of Other Minds'. *The Harvard Review of Philosophy* 25: pp. 51–70. doi: 10.5840/harvardreview20185913.

Sebo, Jeff, and Jason Schukraft. 2021. Don't farm bugs. Aeon. https://aeon.co/essays/on-the-torment-of-insect-minds-and-our-moral-duty-not-to-farm-them

Sebo, Jeff. 2022. *Saving Animals, Saving Ourselves: Why Animals Matter for Pandemics, Climate Change, and Other Catastrophes*. Oxford: Oxford University Press.

Shapiro, Paul. 2018. *Clean Meat: How Growing Meat Without Animals Will Revolutionize Dinner and the World*. New York: Gallery Books.

Shepon, Alon, Gidon Eshel, Elad Noor, and Ron Milo. 2018. 'The Opportunity Cost of Animal Based Diets Exceeds All Food Losses'. *PNAS* 115 (15): pp. 3804–3809. doi: 10.1073/pnas.171382011.

Siipi, Helena. 2008. 'Dimensions of Naturalness'. *Ethics and the Environment* 13 (1): pp. 71–103. doi: 10.2979/ETE.2008.13.1.71.

Siipi, Helena. 2013. 'Is Natural Food Healthy?' *Journal of Agricultural and Environmental Ethics* 26 (4): pp. 797–812. doi: 10.1007/s10806-012-9406-y.

Sinclair, Rebekah. 2016. 'The Sexual Politics of Meatless Meat: (In)edible Others and the Myth of Flesh Without Sacrifice'. In *The Future of Meat Without Animals*, edited by Brianne Donaldson and Christopher Carter, pp. 229–248. London: Rowman & Littlefield International.

Singer, Peter. 1975. *Animal Liberation: A New Ethics for Our Treatment of Animals*. New York: HarperCollins.

Singer, Peter. 2015. *Practical Ethics* (third edition). Cambridge: Cambridge University Press.

Specht, Liz. n.d. 'The Science of Fermentation'. Good Food Institute. https://gfi.org/science/the-science-of-fermentation/.

Springmann, Marco, H. Charles J. Godfray, Mike Rayner, and Peter Scarborough. 2016. 'Analysis and Valuation of the Health and Climate Change Cobenefits of Dietary Change'. *PNAS* 113 (15): pp. 4146–4151. doi: 10.1073/pnas.1523119113.

Stanescu, James K. 2018. 'Matter'. In *Critical Terms for Animal Studies*, edited by Lori Gruen, pp. 222–233. Chicago: University of Chicago Press.

Stănescu, Vasile. 2018. '"White Power Milk": Milk, Dietary Racism, and the "Alt-Right"'. *Animal Studies Journal* 7 (2): pp. 103–128. https://ro.uow.edu.au/asj/vol7/iss2/7.

Stephens, Neil. 2010. 'In Vitro Meat: Zombies on the Menu'. *SCRIPTed* 7 (2): pp. 394–401. doi: 10.2966/scrip.070210.394.

Stephens, Neil. 2013. 'Growing Meat in Laboratories: The Promise, Ontology, and Ethical Boundary-Work of Using Muscle Cells to Make Food'. *Configurations* 21 (2): pp. 159–181. doi: 10.1353/con.2013.0013.

Stephens, Neil, Alexandra E. Sexton, and Clemens Driessen. 2019. 'Making Sense of Making Meat: Key Moments in the First 20 Years of Tissue Engineering Muscle to Make Food'. *Frontiers in Sustainable Food Systems* 3 (45): pp. 1–16. doi: 10.3389/fsufs.2019.00045.

Swift, Adam. 2013. *Political Philosophy: A Beginners' Guide for Students and Politicians* (third edition). Cambridge: Polity.

Taylor, Richard. 1974. *The Joys of Beekeeping*. London: Barrie & Jenkins.

Thompson, Paul B. 2018. 'Farming, the Virtues, and Agrarian Philosophy'. In *The Oxford Handbook of Food Ethics*, edited by Anne Barnhill, Mark Budolfson, and Tyler Doggett, pp. 67–94. Oxford: Oxford University Press. doi: 10.1093/oxfordhb/9780199372263.013.38.

Tong, Q., C. E. Romanini. V. Exadaktylos, C. Bahr, D. Berckmans, H. Bergoug, N. Eterradossi, N. Roulston, R. Verhelst, I. M. McGonnell, and T. Demmers. 2013. 'Embryonic Development and the Physiological Factors that Coordinate Hatching in Domestic Chickens'. *Poultry Science* 92 (3): pp. 620–628. doi: 10.3382/ps.2012-02509.

Treviño, Lindsey S., Elizabeth L. Buckles, and Patricia A. Johnson. 2012. 'Oral Contraceptives Decrease the Prevalence of Ovarian Cancer in the Hen'. *Cancer Prevention Research* 5 (2): pp. 343–349. doi: 10.1158/1940-6207.CAPR-11-0344.

Tschofen, Peter, Inês L. Azevedo, and Nicholas Z. Muller. 2019. 'Fine Particulate Matter Damages and Value Added in the US Economy'. *PNAS* 116 (40): pp. 19857–19862. doi: 10.1073/pnas.1905030116.

Turner, Susan M. 2005. 'Beyond Viande: The Ethics of Faux Flesh, Fake Fur and Thriftshop Leather'. *Between the Species* 13 (5): pp. 1–14. doi: 10.15368/bts.2005v13n5.6.

Urick, M. E., J. R. Giles, and P. A. Johnson. 2009. 'Dietary Aspirin Decreases the Stage of Ovarian Cancer in the Hen'. *Gynecologic Oncology* 112 (1): pp. 166–170. doi: 10.1016/j.ygyno.2008.09.032.

Wadiwel, Dinesh Joseph. 2016. 'Fish and Pain: The Politics of Doubt'. *Animal Sentience* 3 (31): pp. 1–8. doi: 10.51291/2377-7478.1054.

Walsh, John. 2011. 'Sushi: Britain's New National Dish'. The Independent. https://www.independent.co.uk/life-style/food-and-drink/features/sushi-britain-s-new-national-dish-853419.html.

Wayne, Katherine. 2013. 'Permissible Use and Interdependence: Against Principled Veganism'. *Journal of Applied Philosophy* 30 (2): pp. 160–175. doi: 10.1111/japp.12010.

Weele, Cor Van der, and Clemens Driessen. 2013. 'Emerging Profiles for Cultured Meat; Ethics Through and As Design'. *Animals* 3 (3): pp. 647–662. doi: 10.3390/ani3030647.

Wellman, Christopher Heath. 2008. 'Immigration and Freedom of Association'. *Ethics* 119 (1): pp. 109–141. doi: 10.1086/592311.

Whyte, Kyle Powys. 2018. 'Food Sovereignty, Justice, and Indigenous Peoples: An Essay on Settler Colonialism and Collective Continuance'. In *The Oxford Handbook of Food Ethics*, edited by Anne Barnhill, Mark Budolfson, and Tyler Doggett, pp. 345–366. Oxford: Oxford University Press. doi: 10.1093/oxfordhb/9780199372263.013.34.

Wilbanks, Rebecca. 2017. 'Real Vegan Cheese and the Artistic Critique of Biotechnology'. *Engaging Science, Technology, and Society* 3: pp. 180–205. doi: 10.17351/ests2017.53.

Wisnewski, J. Jeremy. 2014. 'Cannibalism'. In *Encyclopedia of Food and Agricultural Ethics*, edited by David M. Kaplan, pp. 279–286. Dordrecht: Springer. doi: 10.1007/978-94-007-0929-4_28.

Wolf, Susan. 2018. 'The Ethics of Being a Foodie'. In *The Oxford Handbook of Food Ethics*, edited by Anne Barnhill, Mark Budolfson, and Tyler Doggett, pp. 722–738. Oxford: Oxford University Press. doi: 10.1093/oxfordhb/9780199372263.013.36.

Woll, Silvia. 2019. 'On Visions and Promises—Ethical Aspects of In Vitro Meat'. *Emerging Topics in the Life Sciences* 3: pp. 753–758. doi: 10.1042/ETLS20190108.

Wright, Laura. 2015. *The Vegan Studies Project: Food, Animals, and Gender in the Age of Terror*. Athens, GA: University of Georgia Press.

Wurgaft, Benjamin Aldes. 2020. *Meat Planet: Artificial Flesh and the Future of Food*. Berkeley: University of California Press.

Young, Sharon M., and Daniel C. Benyshek. 2010. 'In Search of Human Placentophagy: A Cross-Cultural Survey of Human Placenta Consumption, Disposal Practices, and Cultural Beliefs'. *Ecology of Food and Nutrition* 49 (6): pp. 467–484. doi: 10.1080/03670244.2010.524106.

Zamir, Tzachi. 2008. *Ethics and the Beast: A Speciesist Argument for Animal Liberation*. Princeton, NJ: Princeton University Press.

Zangwill, Nick. 2021. 'Our Moral Duty to Eat Meat'. *Journal of the American Philosophical Association* 7 (3): pp. 295–311. doi: 10.1017/apa.2020.21.

Zuolo, Federico. 2020. *Animals, Political Liberalism and Public Reason*. Basingstoke: Palgrave Macmillan.

Index

abolitionism and extinctionism
  about, 3–4
  activism, 3, 178, 179, 182–183
  animal agriculture, 3, 5, 22, 36–39, 111, 112–113, 115, 137–138
  animal rights, 'old', 3–6, 19, 22, 36–39, 113, 115, 158, 186–187
  veganism, 3–6, 36, 39, 117, 186–187
Abrell, Elan, 116 n.2, 117, 120, 132–133
activism
  about, 160–161, 178–183
  abolitionism, 3–4, 178–179, 182–183
  animal rights, 20–21, 35–36, 179, 182–183
  cellular agriculture, 181
  food justice, 11–14, 31, 33
  ideal and non-ideal theory, 14–15, 20, 178
  mixed messages on rights-respecting food, 65–66, 173
  plant-based foods, 180–181
  pragmatism, 19–20
  sentience problem, 180
  transition to zoopolis, 14–15, 175–183
  welfarism, 3, 179, 182–183
  See also abolitionism and extinctionism; food justice; sanctuaries
Adams, Carol, 71, 77, 141–142
aesthetics, 26, 29–31, 84–85, 108, 115
  See also good life; taste
agrarianism
  about, 114–115
  animal worker model, 127, 129–130, 189
  good life, 115
  values, 114–115, 119, 121, 127
  See also pastoral farms
agriculture. *See* agrarianism; animal agriculture; arable agriculture; cellular agriculture; food systems; pastoral farms
Alkon, Alison Hope, 12
Alvaro, Carlo, 78 n.15
animal agriculture
  abolitionism, 3–6, 19–20
  animal rights, 'new', 7–11
  animal rights, 'old', 3–7
  vs. arable agriculture, 34–36
  good life, 27–28

  settler colonialism, 103–104
  transition to zoopolis, 14–15, 175–177
  See also abolitionism and extinctionism; animal rights, new and old approaches; pastoral farms; suffering and death
animal-based food system
  harms to human health, 18, 131–132
  vs. non-vegan systems, 18–19
  transition to zoopolis, 14–15, 175–177
  vs. veganism, 2–3, 184
  in the zoopolis, 19–20
  See also animal agriculture; meat culture; slaughter-based meat; suffering and death
animal ethics
  about, 3–6
  animal rights, 3
  balancing of justice/injustice, 33–34, 48, 52–54, 58–59
  liberal approach, 8–11
  non-identity problem, 155
  pragmatism, 19–22
  turn to political philosophy, 4, 7–11, 13–14
  See also moral philosophy; political philosophy; sentience
animal labour. *See* animal worker model, labour rights
animal rights
  about, 1–6, 13–15, 184–189
  anthropocentrism, 11–13, 19
  balancing of justice/injustice, 24, 33–34, 48, 52–54, 187–188
  citizenship, 122, 179
  enforceable rights, 8–11, 40–42, 51
  infringement of rights, 14, 33–34
  liberal approach, 8–11, 19
  negative rights, 61
  place for animals, 113, 117
  relationship-based rights, 14
  respect, 112–113, 117, 128–129
  sentience problem overview, 2–3, 40–45, 61, 179–180
  in the zoopolis, 113, 184–189
  See also sentience; zoopolis (animal-rights-respecting state)

animal rights, interest-based rights
　about, 13–14, 44, 54–59
　agency and choice, 13–14, 67
　attractive vs. aversive experiences, 43–45, 54–57, 60
　interest in continued life, 55–59
　invertebrates, 54–59
　negative rights, 57–59, 61
　positive rights, 14, 58
　precautionary principle, 48–50, 52–55
　sentience categories/concepts, 42–47, 54–55
　suffering and death, 13–14, 55–59
animal rights, new and old approaches
　about, 3–11, 137–138, 179
　community members, 14, 37–38, 128–129, 137–140, 143, 179
　labour rights, 121–125
　new approaches, 4–11, 42 n.1, 58 n.18, 112–113, 143, 179, 184–189
　old approaches, 3–7, 42 n.1, 58 n.18
　political rights, 148, 179
　See also abolitionism and extinctionism; animal worker model; mail-order cells (MOC); pig in the backyard (PIB); zoopolis (animal-rights-respecting state)
animal worker model
　about, 16, 116, 135, 143
　community members, 128–129, 179
　corpse farming, 128–129
　democracies, 129–131, 147–148
　environmental impacts, 131–133
　FairEggs example, 143–150
　homes, 37–38, 125, 126–128, 132
　ideal and non-ideal theory, 14–15, 134–135
　just food systems, 135, 178, 189
　with MOC and PIB models, 126–128
　as not generalizable, 129–130
　objections to, 128–135
　power decentralization, 127
　respect, 128–129
　revenue from, 127, 133
　source cells problem, 16, 119, 124–127, 132–135, 189
　state role, 126–127, 134, 161
　value conflicts, 130–132
　vs. veganism, 130–132
　zoonotic risks, 131–132
　in the zoopolis, 133–135, 189
　See also eggs, egg farms
animal worker model, labour rights
　about, 121–125, 143
　choice of work, 123, 143, 148–149, 153, 164

citizenship, 122, 179
　FairEggs example, 143–150
　good work, 133–134
　healthcare, 145, 154–157
　labour union, 123–124, 143, 146–148
　political representation, 144–150
　remuneration, 123–125, 143, 145–148
　retirement, 123, 125, 128–129, 143–145, 157
Ankeny, Rachel, 105
apiculture. *See* bees and bee products
arable agriculture
　about, 34–36
　animal rights, 19, 35–36
　harm to animals, 34–36, 38–39, 48–49, 169, 184–187, 189 n.2
　land availability, 32, 168
　vs. non-vegan food systems, 36, 169, 184–187
　pest control, 34, 40, 48–49
　sentience problem, 48–49
　in the zoopolis, 34–36, 169, 182–183, 190
authenticity, 115–116, 119, 131–132

backyard chickens. *See* eggs, backyard chickens
Barnhill, Anne, 167–170
beef. *See* cultivated meat; meat; meat culture; plant-based meat; slaughter-based meat
bees and bee products
　about, 59–61
　apiculture, 54, 59–61, 109, 117, 188
　balancing of justice/injustice, 54, 59–61, 187–188
　bee products, 40–41, 54, 60–61, 110, 180–181
　cellular agriculture, 89, 110
　good life, 60, 186, 188
　sentience problem, 40–42, 54, 59–61, 187, 188
　suffering and death, 59–61
Belshaw, Chris, 29–30
biopsies. *See* cellular agriculture, source cells problem
Birch, Jonathan, 54
birds, 136, 190
　See also chickens
bivalves
　cockles, 24, 41, 163–164
　environmental impacts, 51–52, 168, 174
　as food, 41, 51–52, 180
　mussels, 41, 161
　oysters, 2–3, 26, 41, 51–52, 174, 180, 182–183, 186–187

bivalves (*Continued*)
  Probably Non-Sentient, 51–52, 161, 174, 180
  See also invertebrates
Blattner, Charlotte, 153–154
Bobier, Christopher, 35
Bonotti, Matteo, 167–170
Bray, Heather, 105
breastmilk. *See* motherhood and pregnancy
Bruckner, Donald, 35

cannibalism, 95–98, 140–141
  See also placentophagy
carnism and carnophallogocentrism, 68
  See also meat culture
cats and dogs, 36–37, 95
  See also companion animals
cellular agriculture
  about, 3, 16, 88–89, 110
  absence of animals, 99–100, 104, 108–110, 112, 120–121, 188–189
  activism, 181
  animal rights, 112–113
  availability, 87–89, 99, 172–173
  cannibalism, 95–98, 140–141
  cultivated meat overview, 87–89
  false-hierarchy objection, 95–98, 100
  food justice, 134–135, 163, 186–187
  healthfulness, 131–132
  industries, 87–91, 98, 99–100, 110, 132–133, 165
  just food systems, 98, 111, 166–167, 181, 188–189
  non-food-related products, 191–192
  power relations, 113–114, 121
  products, 88–89, 96, 97–100, 104, 105–106, 110, 191
  taste, 87–88
  in the zoopolis, 3, 90, 110, 188–189
  See also cultivated eggs; cultivated meat; cultivated milk; dairy; genetic engineering; precision fermentation
cellular agriculture, source cells problem
  about, 16, 125–126, 132–133, 189
  animal worker model, 119, 124–127, 132–135, 189
  biopsies, 126, 132–133, 136
  immortal cell lines, 120, 129–130, 132–133, 135
  MOC and PIB models, 117, 189
  respect for animals, 112–113, 117
  slaughter-based meat, 119, 125–126
  umbilical cords, 126
  See also mail-order cells (MOC); pig in the backyard (PIB)
cellular agriculture, technologies
  collagen, 89–91, 99
  FBS (foetal bovine serum), 87–88, 90–92, 172–173, 181
  growth media, 90
  haem (or heme), 63, 84, 88–89, 99, 109
  source cells problem, 91, 99, 119, 125–126, 132–133, 135
cephalopods (octopuses, squids), 40–42, 44, 49–50
chickens
  animal worker model, 127, 129–130, 189
  broodiness, 152–154
  as companion animals, 138–139
  ovarian cancer, 154–157
  PIB model, 117
  reproduction and breeding, 139, 141–143, 151–152, 154–157
  sanctuaries, 182–183
  source cells problem, 117, 125–126
  See also eggs
chimpanzees, 95, 129–130
class, social, 24, 68–71
  See also marginalized communities
clothing, 191–192
Cochrane, Alasdair, 7, 123, 145, 148, 166
cockles. *See* bivalves
collagen, 89–91, 99
companion animals
  animal rights, 14
  backyard chickens, 137–142, 144, 157–158
  community membership, 122, 143, 179, 182
  extinctionism, 22, 36–37
  food for, 37, 97–98 n.12
  positive rights, 14, 58
  source cells problem, 95, 116–118, 135
  in the zoopolis, 22, 36–37, 137
corpses
  corpse farming, 128–129, 190
  respect, 74 n.12, 128–129, 190
  See also roadkill
cows. *See* cultivated milk; dairy; milk
crustaceans
  decapods (shrimps and prawns), 40–41, 49–51, 59, 187
  fishing and farming of, 47, 49–51

as food, 40–41, 47, 49–51
lobsters, 26, 41, 43, 49–50, 53 n.14
Probably Sentient, 49–51, 53 n.14
sentience problem, 49–50, 53 n.14, 59, 187
in the zoopolis, 49–50, 187
See also fish and 'sea food'
cultivated eggs, 110, 125–126, 158
cultivated meat
about, 16, 87–89, 111–112
animal ingredients, 90–92
animal rights, 90, 92–94, 112–113
animal-worker model, 125–126, 131–132
authenticity, 115–116, 119
availability, 87–88, 99, 100, 117, 172–173
cannibalism, 95–98, 140–141
ethical concerns, 87–88, 92–98
false-hierarchy objection, 95–98, 100
historical injustices, 90, 92–94
imitation of 'real' meat, 91, 188–189
industries, 87–91, 117, 132–133
just food systems, 111–112, 166–167, 188–189
in the zoopolis, 90, 110, 166–167
See also cellular agriculture
cultivated meat, technologies
about, 87–89
animal cells as inputs, 90–92, 110, 119, 125–126, 189
collagen, 89–91, 99
haem (or heme), 63, 84, 88–89, 99, 109
human cells as inputs, 96–98
cultivated meat, utopian visions
about, 111–112
commercialization, 117–118
companion animals, 117
place for animals, 113, 117
power relations, 113–114, 121
respect for animals, 112–113, 117
values of meat, 114–116
vs. veganism, 131–132
See also animal worker model; mail-order cells (MOC); pig in the backyard (PIB)
cultivated milk
about, 99–100, 108
absence of animals, 104, 108–110
animal cells as inputs, 125–126, 189
breastmilk, 98, 100
cheese, 99
ethical concerns, 100–104
as food *vs.* non-food, 100–104
genetic engineering, 104–108

industries, 88–89, 98, 99–100, 108, 174 n.12
risks in, 105–108
in the zoopolis, 3, 182–183, 188–189
See also precision fermentation
customs. *See* food practices

dairy
activism, 103, 182
cellular agriculture, 88–89, 99–100, 104, 105–106
cheese, 85–86, 88–89, 99–100, 105–106
environmental impacts, 107–108
injustices, 108–110
milk and milk production, 101–102, 105–106
plant-based products, 85–86, 99–100, 180–181, 189
rBGH (bovine growth hormone), 105–106 n.19
in the zoopolis, 85–86, 103–104, 108–110, 137, 182, 189
See also cultivated milk; milk; precision fermentation
Datar, Isha, 119–120
death. *See* suffering and death
decapods (shrimps/prawns). *See* crustaceans
Deckha, Maneesha, 103–104
democracy
about, 185
animal citizenship, 35, 122, 143, 147–148, 179
animal worker model, 129–131, 147–148
decision-making, 35, 163
political representation (FairEggs), 147–148
public reason, 166–170
sentientist democracy, 148
social democracy, 165–166, 171
in the zoopolis, 165, 185
See also legal systems; political philosophy
disgust, 78–79, 102
dogs and cats, 36–37, 95
See also companion animals
Donaldson, Sue, and Will Kymlicka
abolitionism and extinctionism, 4–6, 22, 36–37, 113
backyard chickens, 139
chicken foetus rights, 150–152
community membership for animals, 37–38, 145, 149
dependent agency, 147 n.21

Donaldson, Sue, and Will Kymlicka (*Continued*)
  reproductive autonomy, 139, 153
  zoopolis, 2–6, 22, 36–37, 113
Driessen, Clemens, 116–117, 127
Dutkiewicz, Jan, 83, 116 n.2, 117, 120, 132–133

eggs
  about, 16, 136–137
  animal rights, 16, 86, 137–138
  chicken foetus rights, 150–152
  cultivated eggs, 110, 125–126, 158
  good life, 157–158
  just food systems, 174–175, 189
  markets for, 16, 137, 143–144
  mixed messages on rights-respecting food, 65–66, 140, 173
  objectification of, 141–143
  plant-based eggs, 85–86, 110, 137, 158, 180–181, 189
  pragmatic approach, 19–20
  sentience problem, 150–152
  as 'something other than food', 141–143
  source cells problem, 117, 125–126
  wildlife eggs, 136, 190
  in the zoopolis, 110, 137, 157–158, 189
  See also chickens
eggs, backyard chickens
  about, 137–143, 157–158
  animal rights, 137–140, 143–144
  co-living with humans, 137–142, 144, 149–150, 157–158
  vs. egg farms, 150
  eggs as chickens' property, 139–140
  eggs as food, 138–143
  objections to, 138–143
eggs, egg farms
  about, 143–152, 157–158
  activism, 182
  animal worker model, 143–150, 153, 154–158, 172
  vs. backyard chickens, 150
  broodiness, 152–154
  commercialization, 143–144, 146, 154, 156
  destruction of eggs, 138, 144–145, 150–152
  FairEggs example, 143–150, 152, 153–157
  objections to, 150–157
  vs. sanctuaries, 182–183
  selective breeding, 154–157
  suffering and death, 144–145, 149–150
  in the zoopolis, 137, 157–158, 182, 189

Elder, Max, 51
employment. *See* good work
environmental concerns
  animal worker model, 131–133
  fish and 'sea food', 50–52, 168, 174
  food justice, 11–13, 168
  land ethics, 30
  non-vegan food systems, 168
  pastoral farms, 108, 115, 132, 168
  plant-based systems, 18, 81, 82–83, 168, 188
  processed foods, 82–83
  public reason, 168
  slaughter-based systems, 80–83
  technologies, 105–108
ethics. *See* animal ethics; animal rights, interest-based rights; political philosophy
Evans, Matthew, 20–22, 28, 31–32
expected utility/expected value, 52–54
extinctionism. *See* abolitionism and extinctionism

Fabre, Cecile, 46
FairEggs, 143–150, 152, 153–157
  See also eggs, egg farms
Fairlie, Simon, 117
false-hierarchy objection, 95–98, 100
farms. *See* agrarianism; animal worker model; pastoral farms
FBS (foetal bovine serum), 87–88, 90, 91–92, 172–173, 181
Fearnley-Whittingstall, Hugh, 140–141
fermentation, 99–100
  See also precision fermentation
Firth, Henry, 93
Fischer, Bob, 72
fish and 'sea food'
  cellular agriculture, 129–130
  food justice, 12–13, 49–50
  food practices, 9, 24, 49–50, 69
  jellyfish, 44–45, 51, 180, 187
  sentience problem, 44–45, 47, 49–51, 180, 187
  suffering and death, 9, 29
  in the zoopolis, 29 n.7, 190
  See also bivalves; cephalopods (octopuses, squids); crustaceans
foodies
  authenticity, 84, 115–116, 119, 131–132
  cellular agriculture, 95, 127
  cultural identity, 49–50, 64

good life, 26
locavores, 115, 119, 127
pastoral farms, 115, 127
plant-based meats, 64
veganism as perceived threat to, 25–26
See also taste
food justice
  about, 11–13, 31–34, 186–187
  access to food, 11, 13, 31–32, 52, 163, 167–168, 186–187
  activism, 11–13, 31, 33
  animal-based systems, 78, 132, 167–168, 186–187
  animal rights, 11–13, 19, 31, 33–34, 49–50, 163–164, 186–188
  anthropocentrism, 11–13, 19
  balancing of justice/injustice, 33–34, 52–54, 187–188
  cellular agriculture, 113–114, 134–135, 163, 186–187
  defined, 12–13
  good life, 11, 49–50, 167–170
  good work, 11, 13, 32–33, 49–50
  hunger as injustice, 31–34, 163
  just food systems, 167–170, 186–187
  liberal approach, 19, 31–34, 163–164
  marginalized people, 11–12, 31–34, 49–50, 170, 188
  non-vegan food systems, 78, 132, 163, 167–170
  public reason, 167–170
  state role, 163, 167–170, 186–188
  veganism as perceived threat to, 31–34, 38–39, 186–187
  in the zoopolis, 19, 34, 52, 186–188
food practices
  about, 23–25
  changes in, 69–71, 73, 77
  culturally appropriate foods, 52 n.13
  good life, 22–26, 31, 60, 78–79
  identity, 23–25, 49–50, 60, 157–158
  liberal approaches, 22–23, 78–79
  meat culture, 68–71
  special occasions, 23–24, 64, 66, 157–158
  veganism as perceived threat, 22–25, 31
  See also religion and food
food systems
  about, 1–3, 16–17, 184–189
  anthropocentrism, 11–13, 19
  changes in, 69–71, 73, 77
  public reason, 166–170

  transition to zoopolis, 14–15, 175–177, 184–189
  See also food justice; just food systems; non-vegan food systems; veganism

Gaard, Greta, 101, 103
Garner, Robert, 5–7, 19–20, 36–37
gender/sex
  chicken foetus rights, 150–152
  feminized protein, 104, 141
  gender identity and food, 24–25
  meat culture, 71
  women as 'absent referent', 71
  See also motherhood and pregnancy
genetically modified organisms (GMOs), 88–89, 105–108, 115, 128–129
genetic engineering
  cultivated milk, 104–108
  genetically modified yeast, 63, 84, 88–89, 99, 106
  genetic editing, 106
  naturalness, 104–105
  precision fermentation, 99–100, 104–108
  rBGH (bovine growth hormone), 105–106 n.19
  risks, 105–108
  in the zoopolis, 105, 107–108
good life
  about, 22–23, 185–189
  diverse concepts, 22–23, 78–79, 186–187
  food justice, 11–13
  food systems, 166–170
  just food systems, 23, 84, 166–170, 185, 186
  liberal approaches, 9–10, 22–23, 166, 185
  meanings in, 22–23, 60
  meat culture, 64, 68–71, 76–78
  naturalness, 84–85
  non-vegan food systems, 19, 27, 169–170
  pastoral farms, 28–31, 108–110, 115, 157–158
  public reason, 166–170
  taste, 25–27, 31, 78
  veganism as perceived threat, 22–23, 31, 32–33, 38–39, 165
  See also food practices; good work; taste
good work
  animal worker model, 123, 127, 133–134, 143, 148–149, 153, 164
  food justice, 11–13, 32–33, 49–50

good work (*Continued*)
    good life, 22, 27–28, 157–158, 164
    non-vegan food systems, 27–28
    pastoral farms, 27–28, 32–33, 115, 127
    transition to zoopolis, 14–15, 164, 175–177
    veganism as perceived threat, 21–23, 27–28, 31, 32–33

Hadley, John, 55, 146 n.20
haem (or heme), 63, 84, 88–89, 99, 109
hate speech, 75–77
Hayek, Matthew, 83
humans
    animal equality to, 95
    anthropocentrism, 11–13, 19
    cannibalism, 95–98 n.12, 140–141
    false-hierarchy objection, 95–98, 100
    imitation of body parts or products, 67, 91, 142
    meat culture, 68–71
    as mere resources, 91–92
    moral horror, 192
    objectification of, 141–143
    source cell problem for cultivated flesh, 96–98
    speciesism, 71, 103
    See also food justice; food practices; good life; good work; meat culture; zoopolis (animal-rights-respecting state)
humans, health
    dietary needs, 31–32
    lactose intolerance, 103–104
    milk as food *vs.* non-food, 100–104, 108
    ultra-processed foods, 80–82
    zoonotic risks, 18, 70, 72, 82, 120–121, 131–132
    See also food justice; motherhood and pregnancy
hunting and trapping, 34–35, 95, 97, 103, 176, 186

ideal and non-ideal theory
    about, 14–15, 134, 184–189
    animal worker model, 134–135
    MOC model, 120
    pragmatism, 19–20, 178
    transition to zoopolis, 14–15, 172, 175–178, 184–189
    in the zoopolis, 14–15, 18, 160–161, 172, 183–186, 192
indigenous peoples, 70, 96–97, 103–104
inequalities. *See* food justice; marginalized communities

insects
    cellular agriculture, 89, 191
    farming, 59–61, 188
    as food, 40–41, 89, 190
    products, 41, 191
    sentience problem, 2–3, 15–16, 40–42, 46, 52, 59, 187–188
    suffering and death, 48–49, 55, 58
    in the zoopolis, 187
    See also bees and bee products; invertebrates
interest-based rights. *See* animal rights, interest-based rights
invertebrates
    about, 40–42
    animal ethics, 46, 53
    balancing of justice/injustice, 48, 52–54, 58–59, 187–188
    as food, 40–41, 44–45, 48, 52–54, 89, 180, 187, 190
    interest-based rights, 55–59
    just food systems, 19, 40–42, 180, 187–188
    negative rights, 57–59, 61
    sentience problem, 15–16, 40–42, 45–46, 48, 52–57, 59, 180
    suffering and death, 55–57
    in the zoopolis, 61, 180, 187–188
    See also bees and bee products; bivalves; cephalopods (octopuses, squids); crustaceans; insects; jellyfish

Jacquet, Jennifer, 51
jellyfish, 44–45, 51, 180, 187
just food systems
    about, 1–3, 160–162, 184–189
    animal rights, 1–6, 40, 41–42, 135, 170–172, 184–189, 192
    cellular agriculture, 3, 98, 111, 188–189
    complexities in, 40, 42, 160–161, 183, 184–186
    good life, 164–167, 169–170, 185, 186
    justice, 192
    liberal approaches, 162–166
    non-vegan systems, 2–3, 39, 163, 166–172, 184
    pastoral farms, 114–116, 163, 168, 178, 189
    public reason as justification, 166–170
    research needed, 183–185
    sentience problem, 2–3, 40–42, 61, 179–180, 187–188
    state role, 161–167, 170
    in the zoopolis, 1–3, 39, 185–189
    See also food justice; zoopolis (animal-rights-respecting state)

Index

justice
  about, 7–11, 184–189
  animal rights, 8–11, 13–15, 184–189
  balancing of justice/injustice, 24, 33–34, 48, 52–54, 187–188
  concept *vs.* conceptions, 7–8, 10
  diverse conceptions of justice, 7–8, 10
  enforceable rights, 8–11, 40–42, 51
  ideal and non-ideal theory, 14–15
  liberal approach, 19
  negative rights, 57–59, 61
  public reason, 169–170
  Rawlsian approach, 7–8, 169–170
  See also food justice; ideal and non-ideal theory; liberalism

Kazez, Jean, 20, 26
Kymlicka, Will. *See* Donaldson, Sue, and Will Kymlicka

labour. *See* animal worker model, labour rights; good work
land
  about, 28–31
  nationalism and landscape, 30–31
  rural visions, 28–31, 108, 115
  veganism as perceived threat, 23, 30–31
  See also environmental concerns; nature and naturalness; pastoral farms
legal systems
  animal political rights, 179
  anti-cruelty laws, 20
  coercive measures, 8, 50, 51, 70, 78–79, 166–167
  enforceable rights, 8–11, 40–42, 51
  hate speech, 75–77
  negative rights, 57–59, 61
  public reason as justification, 166, 170
  sentience problem, 51, 179–180, 187
  See also democracy; liberalism; political philosophy
Leopold, Aldo, 30, 114–115
liberalism
  about, 160–161, 183, 184–189
  animal rights, 8–11, 19, 160–161, 166
  democracies, 163, 165, 185
  diverse concepts of the good, 19, 22–23, 78–79, 130–132, 186
  food justice, 11–13, 19, 162–163, 167–170
  good life, 19, 164, 167, 169–170, 186
  minarchist liberals, 162–164
  non-vegan food systems, 19, 22–23
  perfectionist liberals, 164–166
  public reason, 166–172
  See also democracy; legal systems; political philosophy
lobsters. *See* crustaceans
locavores, 115, 127
logic of the larder, 155

mail-order cells (MOC)
  about, 112, 119–121
  absence of animals, 120–121
  animal worker model, 126–128
  ideal and non-ideal theory, 120
  source cells problem, 120, 125–126, 132–133, 135, 189
  in the zoopolis, 116, 121, 189
  See also cellular agriculture, source cells problem
marginalized communities
  balancing of justice/injustice, 52–54, 187–188
  culturally appropriate food, 52 n.13, 96–97
  food justice, 11–13, 31–34, 49–50, 113–114, 170, 188
  indigenous peoples, 70, 96–97, 103–104
  meat culture, 68–71
  milk consumption, 103–104
  racialized people, 103–104
  veganism as perceived threat, 32–33, 114
  in the zoopolis, 103–104, 175, 185
  See also food justice
marine animals. *See* fish and 'sea food'
McGeown, Richard, 87–88, 92
McMahan, Jeff, 128
meat
  about, 62–64
  healthfulness, 78, 131–132
  naturalness, 84–85
  terminology, 62–64
  in the zoopolis, 3
  See also cultivated meat; meat culture; plant-based meat; slaughter-based meat
meat culture
  about, 68–73
  cannibalism, 95–98, 140–141
  carnophallogocentrism, 68
  changes to culture, 70–71, 73, 77
  cultural meanings, 68–71
  good life, 25–26, 68–71, 76–78
  mixed messages on rights-respecting food, 65–66, 173
  plant-based meat as reaffirming of, 68–71

milk
  ethical concerns, 100–104
  as food *vs.* non-food, 100–104
  historical background, 103–104
  lactose intolerance, 103–104
  marginalized communities, 103–104
  pastoral farms, 108–110
  plant-based milk, 85–86, 107–110, 180–181, 188–189
  pragmatic approach, 19–20, 175–177
  in the zoopolis, 3, 103–104, 107–110, 182
  See also cultivated milk; dairy
Mill, John Stuart, 8–9, 162–163
Miller, John, 88 n.1
minorities. *See* marginalized communities
mixed messages on rights-respecting food, 65–66, 140, 173
MOC. *See* mail-order cells (MOC)
Moore, A. E., 72
moral philosophy, 3–4
  See also animal ethics; animal rights, interest-based rights; political philosophy
motherhood and pregnancy
  breastmilk, 67, 91, 98, 100, 101–102
  dietary needs, 31
  placentophagy, 96, 140–142
mussels. *See* bivalves

nation and nationalism, 24, 30, 69, 115
nature and naturalness
  about, 84–85
  authenticity, 84, 115–116, 119, 131–132
  etymology, 30
  genetic engineering, 104–105
  processed food, 84–85
  rural visions, 28–31, 108, 115
  veganism as perceived threat, 28–31
  in the zoopolis, 85, 105, 108
negative rights, 41–42, 57–59, 61
new omnivorism, 34–35, 173–174 n.10, 189
  See also animal-based food system
new rights. *See* animal rights, new and old approaches
non-ideal theory. *See* ideal and non-ideal theory
non-sentient animals. *See* sentience, Probably Non-Sentient
non-vegan food systems
  about, 1–3, 6, 18–19, 184–189
  animal rights, 1–3, 18–19, 170–172, 184–189
  vs. arable agriculture, 36, 169, 184–187
  author's position, 173–175, 184–189

food justice, 11–13, 78, 132, 163, 167–170
good life, 18–19, 22–23, 27, 28, 169–170
ideal and non-ideal theory, 14–15, 20
just food systems, 39, 163, 166–172
liberal approach, 18–19, 22–23, 31, 166–171
public reason, 166–171
sentience overview, 2–3
terminology, 4
transition to zoopolis, 14–15, 19–20, 175–177
vs. veganism, 4–6, 18–22, 27–28, 169
in the zoopolis, 6, 18–19, 21, 34, 36, 175, 183, 184–189
Nozick, Robert, 10, 162 n.2
Nussbaum, Martha, 42 n.2, 164–165

octopuses and squids, 40–42, 44, 49–50
old rights. *See* animal rights, new and old approaches
omnivores, new, 34–35, 173–174, 189
  See also animal-based food system
ovarian cancer in chickens, 154–157
oysters. *See* bivalves
Ozturk, Burkay, 72

pastoral farms
  about, 28–31, 114–116
  agrarianism, 114–115, 119
  animal rights, 28–29, 34–35, 112–113, 115, 149–150
  animal worker model, 125–127, 129–130, 135, 143, 189
  cellular agriculture, 121, 134–135
  dairy farms, 108–110
  good life, 27–28, 115, 157–158
  good work, 27–28, 32–33, 115, 127
  just food systems, 116, 163, 168, 178, 189
  naturalness, 84–85
  revenue streams, 133–135
  rural visions, 28–31, 108, 115
  values, 27–29, 108–110, 114–115, 119, 121, 127, 130–132, 157–158
  veganism as perceived threat, 27, 32–35
  in the zoopolis, 108–110, 115, 143, 157–158, 182–183
  See also animal worker model
Pérez, José Luis Rey, 123
philosophy. *See* moral philosophy; political philosophy
pig in the backyard (PIB)
  about, 112, 116–119
  animal worker model, 126–130, 189
  commercialization, 117–118, 133, 135

companion animals, 116–119
source cells problem, 116–118, 125–126, 189
in the zoopolis, 116, 119, 189
placentophagy, 96, 140–142
plant-based eggs, 3, 85–86, 110, 137, 158, 180–181, 188–189
plant-based food system, terminology, 4
See also veganism
plant-based meat
  about, 3, 16, 62–64, 85–86, 188–189
  animal rights, 64, 73, 85
  animals as mere resource, 13, 66–68, 91
  availability, 62–63, 84, 88–89, 99–100, 172–173
  as bad food, 64, 77–86, 188–189
  vs. cellular agriculture, 88–89, 91
  disgust, 78–79, 102
  disrespectful to animals, 64–77, 85–86
  environmental impacts, 82–83, 188
  good life, 64, 78, 188
  healthy/unhealthy, 78–79, 81–82, 130, 188
  indirect harms, 67–71
  just food systems, 64, 174, 180–181, 188
  meat culture, 68–73
  mixed messages on rights-respecting food, 65–66, 173
  naturalness/unnaturalness, 84–85
  as processed food, 79–83
  reaffirming of meat culture, 68–71
  vs. slaughter-based meat, 62, 70–71, 78–79, 82–83
  symbolic disrespect, 73–77
  taste, 62, 76
  terminology, 62–64
  in the zoopolis, 3, 64, 82, 85–86, 180–181, 188
plant-based milk, 85–86, 107–110, 180–181, 188–189
Plato, *Republic*, 7, 14, 184
Plausibly Sentient. *See* sentience, Plausibly Sentient and uncertainties
political philosophy
  about, 4, 10, 160–161, 183, 184–189
  animal rights, 'new', 4–11
  enforceable rights, 8–11, 40–42, 51
  food justice, 11–13, 168–170
  interest-based rights, 13–14
  justice issues, 7–11, 168–170
  political turn in animal ethics, 4, 6–8, 13–14
  public reason, 166–170
  values and rights, 130–132

  See also animal rights; democracy; ideal and non-ideal theory; justice; legal systems; liberalism; pragmatism; Rawls, John
Porcher, Jocelyne, 20, 27–28, 32–33, 114–115
positive rights, 14, 58
Post, Mark, 87–88, 92
power relations
  cellular agriculture, 113–114
  MOC model, 121
  See also legal systems
pragmatism
  arguments for non-vegan food system, 19–22
  ideal and non-ideal theory, 14–15, 19–20, 178
  suffering and death, 19–20
  transition to zoopolis, 14–15, 19–20, 175–183
prawns. *See* crustaceans
precautionary principle, 48–50, 52–54, 53 n.15, 55
precision fermentation
  about, 88–89, 99–100, 110
  absence of animals, 88–89, 99, 100
  brewing technologies, 99–100
  cheeses, 88–89, 99, 105–106
  collagen, 90–91, 99
  dairy products, 88–89, 99, 105–107
  environmental impacts, 107
  ethical challenges, 100
  genetic engineering, 99–100, 104–108
  industries, 88–91, 99, 106–107
  risks, 105–108
  source cells, 88–89, 99–100
  technologies, 99–100, 104–108
  in the zoopolis, 105, 108, 110, 174
pregnancy. *See* motherhood and pregnancy
Probably Sentient. *See* sentience, Probably Sentient
processed food
  about, 79–86
  additives, 80, 89
  environmental impacts, 82–83
  healthfulness, 81–82
  naturalness, 84–85
  NOVA classifications, 80–81
  plant-based dairy, 85–86
  plant-based meat, 79–83
  ultra-processed foods, 80–82
property rights, 33–34, 58 n.17, 113, 122, 139–140, 147–148
public reason, 166–170

racialized people. *See* marginalized communities
raw foodists, 21, 80–82
Rawls, John
  conception of the good, 22, 169–170 n.8
  food systems, 163, 167, 168, 170
  ideal and non-ideal theory, 14–15, 160, 172
  justice concept/conceptions, 7–8, 10
  public reason, 167, 169, 170
Regan, Tom, 3, 25, 42 n.1
religion and food
  about, 25
  culturally appropriate foods, 52 n.13
  good life, 22–23, 25, 60, 157–158, 185
  liberal approach, 22–23, 25, 60, 75, 96–97
  special occasions, 24, 60, 69, 157–158
  veganism as perceived threat, 22–25, 31, 60
rennet, 99, 105–106 n.19
reproduction. *See* chickens; eggs
reproduction, human. *See* motherhood and pregnancy
Republic (Plato), 7, 14, 184
respect and disrespect
  animal worker model, 117, 128–129
  corpses, 74 n.12, 128–129, 190
  cultivated meat system, 112–113
  mixed messages on rights-respecting food, 65–66, 140, 173
  symbolic disrespect, 73–77
  See also zoopolis (animal-rights-respecting state)
rights. *See* animal rights; animal rights, interest-based rights; animal rights, new and old approaches
roadkill, 5–6, 35, 174–175, 189, 190
Rousseau, Jean-Jacques, 15
Rützler, Hanni, 87–88

sanctuaries
  vs. farms, 149, 158, 182
  healthcare, 155–156
  microsanctuaries, 155–156, 182
  sanctuary model, 116 n.2, 125, 182
Sandler, Ronald L., 12–13
scavenged food, 174–175
  See also roadkill
Schonwald, Josh, 87–88
sea food. *See* fish and 'sea food'
Sebo, Jeff, 49 n.10, 51, 53 n.14, 129 n.12
seitan, 62, 71, 84
sentience
  about, 2–3, 15–16, 40–48
  attractive *vs.* aversive experiences, 43–45, 54–57, 60
  balancing of justice/injustice, 48, 52–54, 58–59, 187–188
  categories/concepts of sentience, 42–47, 54–55
  chicken foetus rights, 150–151
  invertebrates, 45–46
  just food systems, 40–42, 179–180, 187–188
  legal systems, 179–180
  panpsychism, 47–48
  precautionary principle, 48–50, 52–54, 53 n.15, 55
  research needed, 45–46
  rights-based approach, 2–3, 40–50, 52–59, 61, 187
  sentientism, 40, 42–43
  suffering and death, 41–42, 57–59, 187
  uncertainties, 41–42, 45–47, 179–180, 187
  in the zoopolis, 2–3, 61, 187–188
sentience, Plausibly Sentient and uncertainties
  about, 47–61
  balancing of justice/injustice, 48–50, 52–54, 58–59
  'don't know, don't kill', 48–50
  'don't know, don't worry', 50–52
  expected utility/expected value, 52–54
  'gradations of precaution', 52–54, 53 n.15, 55
  invertebrate farming, 59–61
  rights-based approach, 52–59, 61
  'rights-lite', 54–61
sentience, Probably Non-Sentient
  about, 40–48
  as food, 51–52, 174–175, 179–180
  jellyfish, 44–45, 51, 180, 187
  just food system, 179–180
  problem of non-sentience, 41–47, 187–188
  rights-based approach, 18, 40, 41–42, 61, 187
  See also invertebrates
sentience, Probably Sentient
  about, 46–47
  crustaceans, 49–51, 53 n.14
  precautionary principle, 49–50, 55
  rights-based approach, 49–51
  vertebrates, 44–46, 49–50, 180
settler colonialism, 103–104
sheep, 5–6, 28–29, 109, 136 n.1, 137, 191
shrimps. *See* crustaceans
Sinclair, Rebekah, 86
Singer, Peter, 25–26

slaughter-based meat
  animal as 'absent referent', 71
  environmental impacts, 82–83
  vs. plant-based meat, 62, 68, 70–71, 78–79
  slaughterhouses, 18, 28
  source cells for cultivated meat, 119, 125–126
  in the zoopolis, 90–91
  See also FBS (foetal bovine serum); meat; meat culture
small-scale farms. *See* pastoral farms
social democracy, 165–166, 171
soybeans and soy products, 62, 64, 79–80, 99–100, 106, 131–132, 163–164, 186–187
squids and octopuses, 40–42, 44, 49–50
state. *See* democracy; legal systems; political philosophy; zoopolis (animal-rights-respecting state)
suffering and death
  about, 55–59
  animal worker model, 128–129
  balancing of justice/injustice, 58–61, 187–188
  chickens' cancer example, 154–157
  cruel injustices, 57, 192
  justice issues, 9–11
  pain and sentience, 43, 56–57
  painless death, 19–20, 128, 129
  pragmatic approach, 19–20
  rights-based approach, 13–14, 41–42, 55–59, 187
  sentience problem, 40–42, 54–61, 187
  terminology, 57
  in the zoopolis, 61, 179, 192
  See also sanctuaries; sentience

taste
  about, 25–26
  cultivated meat, 87–88
  disgust, 78–79, 102
  good life, 23–27, 31, 78
  PEF (perfectly ethical food), 26
  plant-based meat, 62, 76
  veganism as perceived threat, 22–26, 31
  See also aesthetics; foodies
Taylor, Richard, 60–61
technologies. *See* cellular agriculture; cellular agriculture, source cells problem; genetic engineering; precision fermentation
tempeh, 62, 79, 80, 99–100
Theasby, Ian, 93

tofu, 62, 64, 79–80
Turner, Susan, 13

unusual eating, 189

veganism
  about, 4, 184–189
  activism, 180–183
  animal rights, 3–6, 18–19, 185–189
  animal worker model, 131–132, 169
  author's position, 173–175, 185–189
  balancing of justice/injustice, 23, 33–34, 52–54, 83, 173–174, 187–188
  environmental impacts, 18, 82–83
  health benefits, 18–19, 82
  ideal *vs.* non-ideal theory, 19–20, 173–174, 178
  moral necessity of, 3–4, 173–175
  vs. non-vegan food systems, 18–22, 27–28
  plant-based meat, 63–64, 82–86, 180–181, 188
  pragmatic approach, 20–22, 178
  sentience problem, 40–43, 48, 51, 61
  terminology, 4
  transition to zoopolis, 14–15, 175–183
  in the zoopolis, 4, 33–34, 38–39, 173–175, 178, 179–189
  See also plant-based meat
veganism, disagreements
  about, 19, 31–34, 38–39, 184, 186–189
  abolitionism and extinctionism, 3–6, 19, 36–39, 186–187
  arable agriculture harms to animals, 34–36, 38–39, 48–49, 169, 182–187, 189 n.2
  environmental impacts, 51–52
  food justice, 19, 31–34, 38–39, 186–187
  good life losses, 22–26, 31–34, 38–39, 184, 186
  good work losses, 27–28, 31, 32–33
  nature connection losses, 28–31
  See also abolitionism and extinctionism
vertebrates
  animal worker model, 129–130
  Probably Sentient, 44–46, 49–50, 180
virtues
  about, 112–113
  agrarian values, 114–116, 132
  meat culture, 72–73
  power relations, 113–114
  respect for animals, 112–113, 117
  reverence, 72

Weele, Cor van der, 116–117, 127
welfarism, 179
  See also sanctuaries
whale meat, 69, 97, 129–130
Whyte, Kyle Powys, 12
wildlife
  eggs as food, 136, 190
  naturalness, 84
  rights, 58–59, 146 n.20
  See also fish and 'sea food'
Wolf, Susan, 25–26
women. *See* gender/sex; humans; motherhood and pregnancy
work. *See* animal worker model; good work

zoonotic risks, 18, 72, 82, 120–121, 131–132
zoopolis (animal-rights-respecting state)
  about, 1–3, 14–15, 160–161, 184–189
  animal rights, 1–3, 113, 169, 184–189
  animals as community members, 22, 36–38, 128–129, 137–140, 143, 179
  arable agriculture, 34–36, 169, 182–183, 190
  author's position, 173–175, 184–189
  cellular agriculture, 90, 110, 166–167, 188–189, 191–192
  cultivated meat, 90, 110, 166–167
  dairy, 85–86, 103–104, 108–110, 137, 182, 189
  defined, 2
  eggs, 110, 137, 157–158, 182, 189
  equality in relationships, 95
  food justice, 11–13, 19, 34, 52, 186–188
  ideal and non-ideal theory, 14–15, 160–161, 172, 183–186, 192
  invertebrates, 61, 180, 187–188
  just food systems, 2–3, 39, 166–171, 185–189
  milk, 103–104, 107–110, 182
  non-food-related products, 191–192
  non-vegan food systems, 18–19, 21, 34, 36, 175
  pastoral farms, 108–110, 115, 143, 157–158, 182–183
  plant-based meat, 64, 82, 85–86, 180–181, 188
  public reason, 166–171
  sentience problem, 1–3, 40–43, 61, 187–188
  state role, 161–162
  suffering and death, 20, 57–59, 61, 179
  transition to, 14–15, 175–177
  veganism, 1–6
  See also animal rights, new and old approaches; animal worker model; just food systems; mail-order cells (MOC); non-vegan food systems; pig in the backyard (PIB); respect and disrespect; veganism